ELECTRICIAN'S GUIDE TO EMERGENCY LIGHTING

Published by The Institution of Engineering and Technology, London, United Kingdom

The Institution of Engineering and Technology is registered as a Charity in England & Wales (no. 211014) and Scotland (no. SCO38698).

The Institution of Engineering and Technology is the new institution formed by the joining together of two great institutions; the IEE (Institution of Electrical Engineers) and the IIE (The Institution of Incorporated Engineers).

© 2009, 2014, 2019 The Institution of Engineering and Technology

First published 2009
Second edition 2014
Third edition 2019

Copies of this publication may be obtained from:
The Institution of Engineering and Technology
PO Box 96, Stevenage, SG1 2SD, UK
Tel: +44 (0)1438 767328
Email: sales@theiet.org
www.theiet.org/wiringbooks

ISBN 978-1-78561-613-6
eISBN 978-1-78561-614-3

Typeset in the UK by the Institution of Engineering and Technology, Stevenage
Printed in the UK by A. McLay and Company Ltd, Cardiff

Contents

Cooperating organisations

The IET acknowledges the invaluable contribution made by the following organisations in the preparation of this Guide.

BEAMA Installations Ltd
P. Sayer IEng MIIE GCGI

British Standards Institution

Certsure
T. Benstead MSc MIET MCIBSE BEd(Hons)
D. Cooney BEng PGCE

Ministry of Housing, Communities and Local Government

Electrical Contractors' Association
R. Giddings IEng MIET ACIBSE

Electrical Contractors' Association of Scotland t/a SELECT
D. Forrester IEng
R. Cairney IEng MIET

Health and Safety Executive
K. Morton CEng MIET

ICEL
C. Watts

Institution of Engineering and Technology
M. Coles BEng(Hons) MIET
G. Gundry MIET

Stroma Certification
J. Peckham MIET LCGI

Welsh Government
C. Blick

Revised, compiled and edited
Eur Ing L. Markwell MSc, BSc(Hons), CEng, MIET, MCIBSE, LCGI,
The Institution of Engineering and Technology, 2019

Acknowledgements

References to British Standards are made with the kind permission of BSI. Complete copies can be obtained by post from:

BSI Customer Services
389 Chiswick High Road
London W4 4AL

For all enquiries contact:

Tel: +44 (0)20 8996 9000
Fax: +44 (0)20 8996 7001

Email: orders@bsi-global.com

References to Building Regulations, Approved Documents and guidance are made with the kind permission of the Ministry of Housing, Communities & Local Government (formerly the Department for Communities and Local Government). Downloads of Approved Documents are available from the Planning Portal: www.planningportal.gov.uk

Preface

The *Electrician's Guide to Emergency Lighting* is one of a number of publications prepared by the IET to provide guidance on electrical installations in buildings. This Guide is concerned with emergency lighting and in particular emergency escape lighting and must be read in conjunction with the relevant legislation and guidance such as Approved Documents B and L in England and Wales and the Building Standards Division (BSD) technical handbooks in Scotland and the relevant British Standards, in particular the BS 5266 series of standards.

Designers and installers should always consult the relevant documents to satisfy themselves of compliance.

It is expected that persons carrying out work in accordance with this Guide will be competent to do so – competence being a statutory requirement of the *Electricity at Work Regulations 1989* for those engaged in electrical work – and a competent person is required to carry out inspections, tests and repairs to emergency lighting systems.

This edition of the *Electrician's Guide to Emergency Lighting* has been revised to align with the new 2016 revision of BS 5266-1 which has introduced new provisions and requirements, and changes from the previous 2011 edition. Designers, installers, dutyholders and persons responsible for premises are advised to make themselves familiar with the new 2016 revision of the standard. This revision took effect from the 31st May 2016.

The emergency lighting installation logbook and all installation certification, testing and maintenance documentation must be retained and made available for review and inspection when required.

Legislation

1

1.1 Introduction

It is necessary to start with a few basic definitions of emergency lighting terms (see diagram below)

▼ **Figure 1.1 Basic definitions**

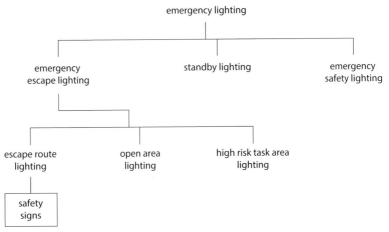

These terms are described more fully in Section 3.3.

Legislation and common law impose duties of care upon those responsible for buildings, generally building owners or tenants and employers but also managers and those responsible for organizing events etc. These persons are generally called 'dutyholders' as they have obligations or duties to provide for the safety of persons using a building or facility. Some persons may not consider that they are dutyholders but if there is an accident the courts will look closely at the circumstances and identify responsibilities.

If precautions and provisions are put in place such that persons are not harmed, problems will not arise. This must be the prime objective, to determine what is necessary for the safety of persons using the facility or building and to implement the requirements. In this context persons include employees, the self-employed and visitors or guests who are in the building or facility and any grounds.

© The Institution of Engineering and Technology

Laws, regulations, standards and guidance documents identify obligations, provide guidance and translate this general duty into specific requirements – numbers, locations, fire performance etc. However, these requirements must be allied with what the designer, installer, operator with their unique knowledge of the facility know to be necessary.

There is considerable legislation and common law that is applicable to emergency lighting. General safety legislation includes:

(a) The Health and Safety at Work etc. Act 1974

(b) The Management of Health and Safety at Work Regulations 1999 (Statutory Instrument 1999 No. 3242)

(c) The Workplace (Health, Safety and Welfare) Regulations 1992 (Statutory Instrument 1992 No. 3004)

(d) The Construction (Design and Management) Regulations 2015 (Statutory Instrument 2015 No.51)

(e) The Health and Safety (Safety Signs and Signals) Regulations 1996 (Statutory Instrument 1996 No. 341).

(f) The Building Regulations 2010 as amended (England and Wales) (Statutory Instrument 2010 No. 2214)

(g) The Building (Scotland) Regulations 2004 as amended (asp 406 for the 2004 Regulations and asp 70 for the 2016 amendment)

Legislation with specific requirements for emergency lighting includes:

(a) The Regulatory Reform (Fire Safety) Order 2005 (Statutory Instrument 2005 No. 1541) and Fire (Scotland) Act 2005 (asp5)

(b) The Cinematograph (Safety) Regulations 1955 (Statutory Instrument 1955 No. 1129) as amended.

These are discussed in more detail below.

1.2 The Health and Safety at Work etc. Act 1974

Together with the common law general duty of care of everyone for his neighbour, the most fundamental legislation with respect to emergency lighting is the Health and Safety at Work etc. Act 1974. It includes comprehensive requirements for health, safety and welfare at work, the control of dangerous substances and certain emissions into the atmosphere. The Act empowers the Secretary of State to make regulations, generally referred to as health and safety regulations, and issue approved codes of practice. The Secretary of State has made much use of this power.

In many respects, the legislation concerned with emergency lighting is similar to that for fire safety. The general duties are given in Section 1:

1. Preliminary

(1)   The provisions of this Part shall have effect with a view to –

    (a)   securing the health, safety and welfare of persons at work; and

    (b)   protecting persons other than persons at work against risks to health or safety arising out of or in connection with the activities of persons at work.

The status of approved codes of practice is given in Section 17 as follows:

17. Use of approved codes of practice in criminal proceedings

(1)   A failure on the part of any person to observe any provision of an approved code of practice shall not of itself render him liable to any civil or criminal proceedings; but where in any criminal proceedings a party is alleged to have committed an offence by reason of a contravention of any requirement or prohibition imposed by or under any such provision as is mentioned in section 16(1) being a provision for which there was an approved code of practice at the time of the alleged contravention, the following subsection shall have effect with respect to that code in relation to those proceedings.

1.3   Construction (Design and Management) Regulations 2015

The designers and constructors of a building etc. are required to consider possible hazards and dangers to health and safety in construction and maintenance, and either design them out or provide safe means for installation, maintenance and repair during the life of a structure. CDM 2015 replaced the Construction (Design and Management) Regulations 2007 (CDM 2007) from 6 April 2015, from this date, the Approved Code of Practice which provided supporting guidance on CDM 2007 was withdrawn.

Under the CDM Regulations the client has specific legal duties and obligations placed upon them . There is a distinction drawn between 'commercial clients' (commercial buildings etc.) and 'domestic clients' (usually domestic dwellings). A commercial client is any individual or organisation that carries out a construction project as part of a business.

Commercial clients have a crucial influence over how their projects are run, including the management of health and safety risks. Whatever the project size, the commercial client has contractual control, appoints designers and contractors, and determines the time and other resources for the project.

For notifiable projects (where planned construction work will last longer than 30 working days and involves more than 20 workers at any one time; or where the work exceeds 500 person days), commercial clients must notify the Health and Safety Executive (HSE) on their standard construction project notification form F10 with details of the project. See HSE document L153 Managing health and safety in construction on the Construction (Design and Management) Regulations 2015 for more detailed information.

▼ **Table 1.1** A brief summary of the main roles and duties under CDM 2015 (from HSE publication L153)

CDM dutyholders: who are they?	Summary of role/main duties
Clients are organisations or individuals for whom a construction project is carried out.	Make suitable arrangements for managing a project. This includes making sure: **(a)** other dutyholders are appointed; and **(b)** sufficient time and resources are allocated. Make sure relevant information is prepared and provided: **(a)** to other dutyholders; **(b)** the principal designer and principal contractor carry out their duties; and **(c)** welfare facilities are provided.
Domestic clients are people who have construction work carried out on their own home, or the home of a family member that is not done as part of a business, whether for profit or not.	Domestic clients are in scope of CDM 2015, but their duties as a client are normally transferred to: **(a)** the contractor, on a single contractor project, or; **(b)** the principal contractor, on a project involving more than one contractor. However, the domestic client can choose to have a written agreement with the principal designer to carry out the client duties.
Designers are those, who as part of a business, prepare or modify designs for a building, product or system relating to construction work.	When preparing or modifying designs, to eliminate, reduce or control foreseeable risks that may arise during: **(a)** construction; and **(b)** the maintenance and use of a building once it is built. Provide information to other members of the project team to help them fulfil their duties.
Principal designers are designers appointed by the client in projects involving more than one contractor. They can be an organisation or an individual with sufficient knowledge, experience and ability to carry out the role.	Plan, manage, monitor and coordinate health and safety in the pre-construction phase of a project. This includes: **(a)** identifying, eliminating or controlling foreseeable risks; and **(b)** ensuring designers carry out their duties. Prepare and provide relevant information to other dutyholders. Provide relevant information to the principal contractor to help them plan, manage, monitor and coordinate health and safety in the construction phase.
Principal contractors are contractors appointed by the client to coordinate the construction phase of a project where it involves more than one contractor.	Plan, manage, monitor and coordinate health and safety in the construction phase of a project. This includes: **(a)** liaising with the client and principal designer; **(b)** preparing the construction phase plan; and **(c)** organizing cooperation between contractors and coordinating their work. Ensure: **(a)** suitable site inductions are provided; **(b)** reasonable steps are taken to prevent unauthorised access; **(c)** workers are consulted and engaged in securing their health and safety; and **(d)** welfare facilities are provided.

CDM dutyholders: who are they?	Summary of role/main duties
Contractors are those who do the actual construction work and can be either an individual or a company.	Plan, manage and monitor construction work under their control so that it is carried out without risks to health and safety. For projects involving more than one contractor, coordinate their activities with others in the project team – in particular, comply with directions given to them by the principal designer or principal contractor. For single-contractor projects, prepare a construction phase plan.
Workers are the people who work for or under the control of contractors on a construction site.	They must: **(a)** be consulted about matters which affect their health, safety and welfare; **(b)** take care of their own health and safety and others who may be affected by their actions; **(c)** report anything they see which is likely to endanger either their own or others' health and safety; and **(d)** cooperate with their employer, fellow workers, contractors and other dutyholders

Note: Table 1.1 is not exhaustive.

1.4 The Management of Health and Safety at Work Regulations 1999 (SI 1999 No. 3242)

The Management of Health and Safety at Work Regulations include a requirement for risk assessment.

3. Risk assessment

Every employer shall make a suitable and sufficient assessment of –

(a) the risks to the health and safety of his employees to which they are exposed whilst they are at work; and

the risks to the health and safety of persons not in his employment arising out of or in connection with the conduct by him of his undertaking.

(b) for the purpose of identifying the measures he needs to take to comply with the requirements and prohibitions imposed upon him by or under the relevant statutory provisions.

This requirement has particular relevance to fire systems, including emergency lighting.

European Council directives require emergency lighting, and the consequence is that many of the health and safety regulations issued by the Secretary of State with the authority of the Health and Safety at Work Act are also implementing European Council directives as well as meeting UK initiatives; see Chapter 2, Building Regulations.

1.5 The Workplace (Health, Safety and Welfare) Regulations 1992 (SI 1992 No. 3004)

In Regulation 8 of the Workplace (Health, Safety and Welfare) Regulations there is a requirement that suitable and sufficient emergency lighting shall be provided in any room in circumstances in which persons at work are specially exposed to danger in the event of failure of artificial lighting. This is a general requirement to provide standby lighting. See Regulation 8 below.

8. Lighting

(1) Every workplace shall have suitable and sufficient lighting.

The lighting mentioned in paragraph (1) shall, so far as is reasonably practicable, be by natural light.

(2) Without prejudice to the generality of paragraph (1), suitable and sufficient emergency lighting shall be provided in any room in circumstances in which persons at work are specially exposed to danger in the event of failure of artificial lighting.

This should also be taken to require suitable emergency lighting on construction sites and anywhere persons are working under artificial light.

1.6 The Health and Safety (Safety Signs and Signals) Regulations 1996 (SI 1996 No. 341)

Where the risk assessment required by the Management of Health and Safety at Work Regulations indicates a need for any safety signs or signals, the Health and Safety (Safety Signs and Signals) Regulations require that suitable signs be installed with, if necessary, a guaranteed emergency electrical supply, detailed in item 8 of Schedule 1, as follows:

Regulation 4 and relevant paragraphs of schedule 1 of the Health and Safety (Safety Signs and Signals) Regulations 1996 are reproduced below.

4. Provision and maintenance of safety signs

(1) Paragraph (4) shall apply if the risk assessment made under paragraph (1) of regulation 3 of the Management of Health and Safety at Work Regulations 1992 indicates that the employer concerned, having adopted all appropriate techniques for collective protection, and measures, methods or procedures used in the organisation of work, cannot avoid or adequately reduce risks to employees except by the provision of appropriate safety signs to warn or instruct, or both, of the nature of those risks and the measures to be taken to protect against them.

(4) Where this paragraph applies, the employer shall (without prejudice to the requirements as to the signs contained in regulation 11(2) of the Offshore Installations (Prevention of Fire and Explosion, and Emergency Response) Regulations 1995) –

    (a) in accordance with the requirements set out in Parts I to VII of Schedule 1, provide and maintain any appropriate safety sign (other than a hand signal or verbal communication) described in those Parts, or ensure such sign is in place; and

    (b) subject to paragraph (5), in accordance with the requirements of Parts I, VIII and IX of Schedule 1, ensure, so far as is reasonably practicable, that any appropriate hand signal or verbal communication described in those Parts is used; and

    (c) provide and maintain any safety sign provided in pursuance of paragraph (6) or ensure such sign is in place.

SCHEDULE 1 – Regulation 4(4) and (5)

PART I

MINIMUM REQUIREMENTS CONCERNING SAFETY SIGNS AND SIGNALS AT WORK

8.　　Signs requiring some form of power must be provided with a guaranteed emergency supply in the event of a power cut, unless the hazard has thereby been eliminated.

9.　　The triggering of an illuminated sign and/or acoustic signal indicates when the required action should start; the sign or signal must be activated for as long as the action requires. Illuminated signs and acoustic signals must be reactivated immediately after use.

10.　　Illuminated signs and acoustic signals must be checked to ensure that they function correctly and that they are effective before they are put into service and subsequently at sufficiently frequent intervals.

The Health and Safety Executive has issued guidance on these regulations in publication L64. This Guide details the recommended signs and signals, including emergency escape and fire-fighting signs. Chapter 7 reproduces many of the recommended signs.

1.7 The Building Regulations 2010 (SI 2010 No. 2214)

The UK Building Regulations have been replaced over recent years by separate local requirements for each of Scotland, England, Wales and N Ireland administered by their governments.

In England, the legislation is Building Regulations 2010 (SI 2010 no 2214) Requirement B1 'Means of warning and escape' of schedule 1 of the Building Regulations is reproduced in Chapter 2.

The Secretary of State issues 'Approved Documents' that offer practical guidance on the requirements of the Building Regulations in England. There is no obligation to adopt any particular solution in an Approved Document, if the designer wishes to meet the requirement in another way. However, all the most common situations are dealt with in a practical way. Approved Document B provides guidance on the requirements of schedule 1 and regulation 7 of the Building Regulations and is discussed in Chapter 2.

Part L of the Building Regulations covers the conservation of fuel and power and as emergency lighting uses electrical energy its consumption will need to be accounted for in building energy design and management systems. Also modifications to lighting systems are a controlled service and need to comply with the requirements of the Building Regulations. The requirements of Part L are discussed further in Chapter 4.

In Scotland the legislation is The Building (Scotland) Regulations 2004 as amended (asp 406). The legislation applies generally to buildings and work carried out in buildings in Scotland with some exemptions. Compliance with the Building Regulations is achieved by fulfilling the requirements of the mandatory building standards and guidance on compliance with these is provided in two technical handbooks produced by the Scottish Government Building Standards Division. There is a handbook for domestic premises and another for non-domestic premises.

1.8 Fire safety and the Regulatory Reform (Fire Safety) Order 2005 (SI 2005 No. 1541)

The Regulatory Reform (Fire Safety) Order 2005 was introduced in October 2006 and was intended to rationalize and simplify much of the legislation concerning fire safety at work in England and Wales; it amended or replaced over 100 pieces of legislation including many local authority acts and the provision of fire certificates. It replaced fire certification under the Fire Precautions Act 1971 with a general duty to those in control of a premises.

The Regulatory Reform (Fire Safety) Order 2005 applies to the majority of premises and workplaces but generally does not apply to dwellings.

The Order requires the responsible person to carry out a fire risk assessment and prepare a policy document for fire safety including emergency lighting. Under this legislation BS 5266 is no longer a rigid requirement, it is for the designer and responsible persons to demonstrate to relevant authorities that adequate emergency lighting – and other necessary precautions – are provided for the safety of occupants and visitors in a premises. Escape procedures must be developed, staff trained, means of escape prepared including escape signs, escape notices, emergency lighting, fire detection and alarm systems installed and fire-fighting equipment positioned as necessary.

Regulation 14 has specific requirements for emergency routes and exits and is reproduced below.

14. Emergency routes and exits

(1) Where necessary in order to safeguard the safety of relevant persons, the responsible person must ensure that routes to emergency exits from premises and the exits themselves are kept clear at all times.

(2) The following requirements must be complied with in respect of premises where necessary (whether due to the features of the premises, the activity carried on there, any hazard present or any other relevant circumstances) in order to safeguard the safety of relevant persons –

(a)–(f) omitted

(g) emergency routes and exits must be indicated by signs; and

(h) emergency routes and exits requiring illumination must be provided with emergency lighting of adequate intensity in the case of failure of their normal lighting.

The Order revoked much legislation, see Table 1.2 below:

▼ **Table 1.2**   Typical examples of principal legislation that were revoked or amended

Instrument	Reference	Extent of revocation
The Fire Certificate (Special Premises) Regulations 1976	SI 1976 No. 2003	The whole Regulations
The Fire Precautions (Workplace) Regulations 1997	SI 1997 No. 1840	The whole Regulations
The Fire Precautions (Workplace) (Amendment) Regulations 1999	SI 1999 No. 1877	The whole Regulations
The Management of Health and Safety at Work Regulations 1999		In regulations 1(2), 3(1), 7(1), 11(1)(a), 11(1)(b), 12(1)(b) the words 'and by Part II of the Fire Precautions (Workplace) Regulations 1997' in each place where they occur.
		In regulation 10(1)(c) the words from 'and the measures' to 'Regulations 1997'.
		In regulation 10(1)(d) the words 'and regulation' to 'Regulations 1997'.
		In regulations 11(2) and 12(2) the words in brackets.
		Regulation 28.

1.9   Fire (Scotland) Act 2005

The Fire (Scotland) Act 2005 covers all aspects of fire safety in Scotland, including the provision of fire and rescue authorities and services, emergency water supplies, the powers of authorities in relation to public safety and enforcement of regulations under the Act and fire safety in buildings.

Requirements are similar to those contained in the Regulatory Reform Fire Safety Order applicable in England and Wales as Part 3 of the Act places duties on employers and persons in control of non-domestic premises in Scotland to ensure the safety of employees and others in respect of harm caused by fire in the workplace. The dutyholder(s) are required to carry out and review assessments for the purpose of identifying risks to the safety of persons in the premises and implement fire safety measures as necessary. This would include for most buildings adequate emergency lighting and safety signs to facilitate the safe escape from the premises in the event of fire.

With specific regard to emergency lighting the Fire Safety (Scotland) Regulations 2006 require in Regulation 13.(g) emergency routes and exits must be indicated by signs; and in (h) emergency routes and exits requiring illumination must be provided with emergency lighting of adequate intensity in the case of failure of their normal lighting. Below is a list of other sources of information which relates to emergency lighting:

(a) The Ministry of Housing, Communities & Local Government (formerly Department for Communities and Local Government) website has advice on the legislation, including premises-specific guidance documents for the Regulatory Reform (Fire Safety) Order 2005.

(b) The Welsh Government website also provides information.

(c) The Scottish Government provides sector specific guidance to help those with responsibilities under the Fire (Scotland) Act 2005 to achieve compliance.

(d) The HSE website has guidance on fire safety in the construction industry.

1.10 The Cinematograph (Safety) Regulations 1955 (SI 1955 No. 1129) as amended and the Cinematograph (Safety) (Scotland) Regulations 1955 (SI 1955 No. 1125) as amended

The regulations require 'safety lighting'. Regulation 16.-(1) states:

Safety lighting

16.-(1) In addition to the general lighting, means of illumination adequate to enable the public to see their way out of the premises without assistance from the general lighting shall be provided –

(a)   in the auditorium and all other parts of the building to which the public are admitted;

(b)   in all passages, courts, ramps and stairways to which the public have access and which lead from the auditorium to outside the premises; and

(c)   for the illumination of all notices indicating exits from any part of the premises to which the public are admitted.

(2) The safety lighting shall be kept on at all times when the public are on the premises except in those parts of the premises which are lit equally well by daylight.

These documents are still current legislation but their requirements would be inadequate to today's standards and more recent legislation includes rather more comprehensive requirements to protect all people using the premises – both staff and the public. Cinemas are included within the scope of BS 5266-1: 2016 (see Clause 9.4).

1.11 Disabled persons

A fire safety risk assessment guide (Means of escape for disabled people) has been published by the government to assist in the safe evacuation of disabled persons from non-residential premises. It advises on the preparation and practice of plans for evacuation of disabled people, including staff, by the use of Personal Emergency Evacuation Plans and plans for visitors. (This document is available for free download from the www.gov.uk website).

Building Regulations

2

2.1 Introduction

2.1.1 The Building Regulations 2010 (England and Wales)

The Building Regulations 2010 (England and Wales), SI 2010 No. 2214, apply to building work as described in regulation 3 of the Building Regulations. In the regulations, building work means:

(a) the erection or extension of a building;

(b) the provision or extension of a controlled service or fitting in or in connection with a building;

(c) the material alteration of a building or a controlled service or fitting;

(d) work relating to a material change of use;

(e) insulation filling of cavity walls; and

(f) work involving the underpinning of a building.

The fire safety requirements of the Building Regulations are given in part B of schedule 1 to the regulations. The particular requirement with respect to emergency lighting is requirement B1 'Means of warning and escape'.

▼ **Table 2.1**   Requirement B1 of schedule 1 to the Building Regulations

Requirement	Limits on application
Means of warning and escape	
B1. The building shall be designed and constructed so that there are appropriate provisions for the early warning of fire, and appropriate means of escape in case of fire from the building to a place of safety outside the building capable of being safely and effectively used at all material times.	Requirement B1 does not apply to any prison provided under Section 33 of the Prisons Act 1952 (power to provide prisons etc.).

Guidance on schedule B1 to the Building Regulations is given in Approved Document B: Fire safety. This Approved Document comes in two volumes:

(1) Volume 1 – dwellinghouses

(2) Volume 2 – buildings other than dwellinghouses.

The guidance given in this chapter is based on the guidance in Volumes 1 and 2 of the Approved Document.

2.1.2 Wales

On 31 December 2011 the power to make building regulations for Wales was transferred to Welsh Ministers. This means Welsh Ministers will make any new building regulations or publish any new building regulations guidance applicable to Wales from that date.

The Building Regulations 2010 and related guidance, including Approved Documents as at that date, will continue to apply in Wales until Welsh Ministers make changes to them. The 2006 version of Approved Document P, although rebranded for use in Wales, still applies at the date of publication of this Guide. The document can be viewed at:

http://gov.wales/docs/desh/publications/170403building-regs-approved-document-p-electrical-safety-en.pdf

Where guidance is reviewed and changes made Welsh Ministers will publish separate approved documents where necessary.

Updated versions of the Welsh Part B (Fire safety) were published in 2014 and 2016 to coincide with the requirements to install automatic fire suppression systems in new and converted residences in Wales. The Approved Documents have been rebranded for use in Wales. The documents can be viewed at:

http://gov.wales/topics/planning/buildingregs/approved-documents/part-b-fire/?lang=en

Approved Document B Volume 1 covers dwellinghouses and Volume 2 covers buildings other than dwellinghouses. Both volumes have five requirements:

- **(1)** warning and escape;
- **(2)** internal fire spread (linings);
- **(3)** internal fire spread (structure);
- **(4)** external fire spread; and
- **(5)** access and facilities for the fire service.

2.1.3 Status of Approved Documents

The 'Planning Portal' advises: The Approved Documents are intended to provide guidance for some of the more common building situations. However, there may well be alternative ways of achieving compliance with the requirements. Thus there is no obligation to adopt any particular solution contained in an Approved Document if it is preferred to meet the relevant requirement in some other way.

They are given legal status by the Building Act 1984. Regulation 6 states:

6. Approval of documents for purposes of building regulations

(1) For the purpose of providing practical guidance with respect to the requirements of any provision of building regulations, the Secretary of State or a body designated by him for the purposes of this section may –

    (a) approve and issue any document (whether or not prepared by him or by the body concerned), or

    approve any document issued or proposed to be issued otherwise than by him or by the body concerned,

    (b) if in the opinion of the Secretary of State or, as the case may be, the body concerned the document is suitable for that purpose.

Regulation 7 states:

7. Compliance or non-compliance with approved documents

(1) A failure on the part of a person to comply with an approved document does not of itself render him liable to any civil or criminal proceedings; but if, in any proceedings whether civil or criminal, it is alleged that a person has at any time contravened a provision of building regulations –

    (a) a failure to comply with a document that at that time was approved for the purposes of that provision may be relied upon as tending to establish liability, and

    (b) proof of compliance with such a document may be relied on as tending to negative liability.

(2) **In any proceedings, whether civil or criminal –**

    (a) a document purporting to be a notice issued as mentioned in Section 6(3) above shall be taken to be such a notice unless the contrary is proved, and

    (b) a document that appears to the court to be the approved document to which such a notice refers shall be taken to be that approved document unless the contrary is proved.

2.2 Building (Scotland) Act 2003

This section explains the requirements for electrical installations in Scotland as covered by the Building (Scotland) Act 2003 and associated legislation. Detailed information on the Scottish system including building regulations can be found at the Scottish Government Building Standards Division (BSD) website: www.scotland.gov.uk/bsd.

Building (Scotland) Act 2003

This Act gives Scottish Ministers the power to make building regulations to secure the health, safety, welfare and convenience of persons in or about buildings and of others who may be affected by buildings or matters connected with buildings; to further the conservation of fuel and power; and to further the achievement of sustainable development. The Act also allows Ministers to issue guidance documents in support of these regulations.

The Building (Scotland) Regulations 2004 (as amended)

The Building (Scotland) Regulations 2004 (as amended) are made under the powers of the Building (Scotland) Act 2003 and apply solely in Scotland. The regulations apply to building work such as alterations, extensions, conversions, erections and demolition of buildings and also to the provision of services, fittings and equipment in, or in connection with, buildings except where they are specifically exempted from Regulations 8 to 12 (Schedule 1 to Regulation 3).

The regulations prescribe functional standards for buildings, which can be found in Schedule 5 to Regulation 9. The regulations and functional standards are amended periodically, however, it is the regulations in force at the time of the application for a building warrant that must be complied with.

In Scotland, final responsibility for compliance with building regulations rests with the 'relevant person' (defined in Section 17 of the Act), who will normally be the owner or developer of a building. However, any person carrying out work, including electrical work, has a duty to ensure their work complies with the building regulations.

2.2.1 Technical guidance

The BSD publish two technical handbooks electronically, one for domestic buildings and another for non-domestic buildings, both of which are available to download from https://www.gov.scot/policies/building-standards/monitoring-improving-building-regulations/. The handbooks are arranged in eight sections as below. Section 0 covers general issues and sets out how and when the building regulations apply to buildings and works and the other sections, which each cover a number of related standards, provide guidance on compliance with the functional standards:

(a) Section 0 :General
(b) Section 1: Structure
(c) Section 2: Fire
(d) Section 3: Environment
(e) Section 4: Safety
(f) Section 5: Noise
(g) Section 6: Energy
(h) Section 7: Sustainability

Functional standards of particular relevance to emergency lighting are Section 2: Fire where mandatory standard 2.10 requires *"Every building must be designed and constructed in such a way that in the event of an outbreak of fire within the building, illumination is provided to assist in escape"* and Section 4: Safety where mandatory standard 4.5 requires that *"Every building must be designed and constructed in such a way that the electrical installation does not:*

(a) *threaten the health and safety of the people in, and around ,the building and*
(b) *become a source of fire"*

2.2.2 Procedures

In Scotland, a building warrant is required for building work unless:

(a) it is not a 'building' as defined in the Building (Scotland) Act 2003;

(b) it is exempt from building regulations (Schedule 1 to Regulation 3); and

(c) it does not require a warrant but must still comply with building regulations (Schedule 3 to Regulation 5).

Anyone intending to carry out electrical work that requires a building warrant should note that:

(a) a building warrant must be obtained before work starts;

(b) a completion certificate must be submitted when work is complete; and

(c) a completion certificate must be accepted by the verifier before occupation or use of a new building (or extension) is permitted.

Except where described above a building warrant must be applied for and be granted by the Verifier before work can commence. It is an offence to commence work without a warrant. The application is made to the local authority building standards department (currently only the 32 local authorities have been appointed as Verifiers for their geographical area), who will assess the application and issue the building warrant if proposals are considered to comply with building regulations.

The use of an Approved Certifier of Design or Construction does not remove the need to obtain a building warrant but the certificate they issue must be accepted by the Verifier.

Designers and installers are directed to Section 2 – Fire of the Scottish Building Standards Technical Handbooks for detailed guidance on issues including openings and fire stopping (Standards 2.1, 2.2 and 2.4), escape route lighting (Standard 2.10), fire alarm and detection systems (Standard 2.11) and automatic life safety fire suppression systems (Standard 2.15).

If a building warrant has been granted, a Completion Certificate must be submitted to the Verifier once works are complete. This states that works have been completed in accordance with building regulations and the building warrant. If the Verifier, after making their reasonable inquiries, is satisfied, they must accept the Completion Certificate. If it is rejected, the Verifier will identify the reason for rejection.

Where an Approved Certifier of Construction for electrical installations is used, they will issue a Certificate of Construction which the Verifier must accept as proof of compliance for the specific work described. It should be submitted with the Completion Certificate.

2.2.3 Buildings and services exempt from building regulations:

Certain buildings and services are exempt from Regulations 8 to 12 and a building warrant is not required. They are set out in Schedule 1 to Regulation 3 of the Building (Scotland) Regulations 2004 (as amended). The exempt types include:

(a) Type 1 to 3 buildings or work covered by other legislation, for example, Manufacture and Storage of Explosives Regulations, Nuclear Installations Act;

(b) Type 4 protective works, for example building site works;

(c) Type 5 or 6 buildings not frequented by people, for example detached buildings housing fixed plant only requiring intermittent visits;

(d) Type 7 and 8 Agricultural and related buildings, for example commercial greenhouses

(e) Type 9 to 12 Buildings or work that is so specialized that the Buildings Regulations are largely inappropriate, for example, civil engineering works;

(f) Type 13 and 17 to 20 small single storey buildings which do not contain flues, fixed combustion appliances or sanitary facilities, for example, conservatories or porches not exceeding 8 m^2 in area; and

(g) Type 14 to 16 temporary buildings not containing sleeping accommodation such as contractors' site huts.

Certain works subject to building regulations do not require a building warrant but work must still meet the regulations. These are set out in Schedule 3 to Regulation 5 of the Building (Scotland) Regulations 2004 (as amended) and include exceptions where a warrant is required. Further information on this can be obtained from the relevant Verifier.

Electrical installations must comply with Standard 4.5 (electrical safety), which states that every building must be designed and constructed in such a way that the electrical installation does not threaten the health and safety of the people in and around the building, and does not become a source of fire.

There are no significant differences in general installation requirements for carrying out electrical work in Scotland, England and Wales as BS 7671 (as amended) is the recommended means of satisfying building standards requirements. However, Part P electrical self-certification schemes in England and Wales are not applicable in Scotland.

In Scotland, qualified and experienced electricians can certify that their installation work meets the requirements of building regulations under the Certification of Construction (Electrical Installations to BS 7671) scheme approved under Section 7(2) of the Building (Scotland) Act 2003.

2.3 Northern Ireland

The Northern Ireland Building Regulations are legal requirements made by the Department of Finance and Personnel under the Building Regulations (Northern Ireland) Order 1979 (as amended 1990 and 2009) and administered by eleven local councils. The Regulations are intended to ensure the safety, health, welfare and convenience of people in and around buildings.

Under the 1979 Order, the Department is empowered to write Building Regulations for certain matters set out in the Order. The current regulations are the Building Regulations (Northern Ireland) 2012, which came into operation on 31st October 2012.

Building Regulations set requirements and standards for building that can reasonably be attained, having regard for the health, safety, welfare and convenience of people in or around buildings and others affected by buildings or building matters. They also further the conservation of fuel and power, and make provisions for access to buildings.

Building Control is responsible for ensuring that the Building Regulations, a set of construction standards laid down by Parliament, are enforced in your local Council. The standards include requirements on health, structural stability, fire safety, energy conservation and accessibility. These standards are enforced through plan assessment and site inspection by impartial professionals with a thorough knowledge of The Building Regulations and other relevant British Standards, Codes of Practice and guidance.

When a building is completed Building Control will issue a Certificate of Completion once it is satisfied that all necessary Building Regulations have been adhered to.

NI Technical Booklet D 2012 covers structures, Technical Booklet E 2012 deals with fire safety and Booklets F1 2012 and F2 2012 cover conservation of fuel and power in dwellings and buildings other than dwellings respectively.

2.4   Dwellinghouses

A dwellinghouse is generally defined as: A unit of residential accommodation occupied (whether or not as a sole or main residence):

(a) by a single person or by people living together as a family; or
(b) by not more than six residents living together as a single household, including a household where care is provided for residents.

A dwellinghouse does not include a flat or a building containing a flat.

As the scope of Volume 1 is dwellinghouses, and it does not include flats or "houses in multiple occupation" (HMO), there is no requirement for escape lighting but householders may wish to provide some escape lighting in their homes if they wish – perhaps on staircases or near consumer units.

In Scotland, the BSD Technical handbook for domestic buildings guidance is that emergency lighting is not required in a dwelling except buildings considered to be at higher risk and in a building containing flats or maisonettes.

2.5   Buildings other than dwellinghouses

2.5.1   Lighting of escape routes

(Para 5.36 of Approved document B volume 2 in England and para 6.36 of volume 2 in Wales)

Generally all of the building regulations for England, Wales, Scotland and N Ireland require emergency escape lighting in buildings other than dwellinghouses, and they all base their requirements on fire risk assessments and the application of BS 5266. For convenience we will consider the requirements of the English building regulations as examples.

Volume 2 of Approved Document B recommends that all escape routes should have adequate artificial lighting, and that certain escape routes and areas should also have escape lighting to illuminate the escape route should the mains supply fail, see Table 2.2. The lighting to escape stairs should be on a separate circuit from the circuits supplying the lighting to other parts of the escape route. The intention is to provide for a more reliable supply to escape stairs and designers should bear this intention in mind when designing the lighting.

▼ **Table 2.2**   Provisions for escape lighting (Table 9 of App Doc B, Vol. 2)

Purpose group of the building or part of the building		Areas requiring escape lighting
1.	**Residential**	All common escape routes*, except in 2-storey flats
2.	**Office, storage and other non-residential**	a. Underground or windowless accommodation
		b. Stairways in a central core or serving storey(s) more than 18 m above ground level
		c. Internal corridors more than 30 m long
		d. Open-plan areas of more than 60 m^2
3.	**Shop, commercial and car parks**	a. Underground or windowless accommodation
		b. Stairways in a central core or serving storey(s) more than 18 m above ground level
		c. Internal corridors more than 30 m long
		d. Open-plan areas of more than 60 m^2
		e. All escape routes* to which the public are admitted (except in shops of three or fewer storeys with no sales floor more than 280 m^2 provided that the shop is not a restaurant or bar)
4.	**Assembly and recreation**	All escape routes*, and accommodation except for:
		a. accommodation open on one side to view sport or entertainment during normal daylight hours
5.	**Any purpose group**	a. All toilet accommodation with a floor area over 8 m^2
		b. Electricity and generator rooms
		c. Switchroom/battery room for emergency lighting system
		d. Emergency control room

* Including external escape routes.

(Para
5.37 of
Volume 2
in England
and para
6.37 in
Wales)

2.5.2 Exit signs

Volume 2 of Approved Document B recommends that, except within a flat, an exit sign should distinctively and conspicuously mark every escape route. The sign should be of an adequate size complying with the *Health and Safety (Safety Signs and Signals) Regulations 1996*. Escape routes in ordinary use, that is typically the main entrance door of a building, are excluded from this recommendation.

There is a recommendation that, in general, signs should conform to BS 5499-1 *Graphical symbols and signs. Safety signs, including fire safety signs. Specification for geometric shape, colours and layout* (now replaced by BS ISO 3864-1:2011).

Chapter 7 of this publication provides information on the Safety Signs and Signals Regulations and on the British Standard.

(Para
5.38 of
Volume 2
in England
and para
6.38 in
Wales)
(Clause
8.2.2
of BS
5266-1)

2.5.3 Critical electrical circuits

Where it is critical for electrical circuits to be able to continue to function during a fire, such as circuits to luminaires and signs from a central standby supply, Approved Document B calls for 'protected circuits'. A protected circuit for operation of equipment in the event of fire is required by Approved Document B to:

(a) consist of cable that will survive at least 30 minutes when tested to BS EN 50200:2015, Annex E, or an equivalent standard;

(b) follow a route selected to pass only through parts of the building in which the fire risk is negligible; and

(c) be separate (i.e. segregated) from any circuit provided for another purpose.

The Building Regulations add that in large or complex buildings where the fire protection system needs to operate for extended periods, guidance should be sought from BS 5266-1 (and BS 5839-1 and BS 8519:2010 (which superseded BS 7346-6)).

BS 5266-1 describes two levels of fire resistance:

1. standard performance cables and cable systems with an inherently high resistance to attack by fire; and

2. enhanced performance cables and cable systems with an inherently high resistance to attack by fire.

Standard performance cables and cable systems (with high resistance to fire) are ordinarily used and high performance cables and cable systems (with inherently high resistance to fire) are used, for example, when evacuation is staged and or the building is unsprinklered, please see Section 5.3.

Standard cables and cable systems are required generally to have a duration of survival of 60 minutes and enhanced cables or cable systems a duration of survival of 120 minutes. Section 5.3 describes cables and cable systems intended to meet this duration of survival times. Cable systems and cable supports must be designed and securely fixed to prevent collapse or significant distortion of cable containment during elevated temperatures in fire situations. The use of, or requirement for metal clips or containment systems should be considered.

2.5.4 Lighting energy efficiency

Under the requirements of Part L non-domestic lighting in both new buildings or in replacement systems in existing buildings must meet specified energy targets. This can be done by either selecting luminaires that meet the recommended minimum efficacy standards or calculating the overall energy consumption of the installation, including parasitic energy usage, by the Lighting Energy Numerical Indicator (LENI) procedure and comparing with stated maximum energy per square metre per year limits (see the Part L Non-Domestic Building Services Compliance Guide for further details).

In Scotland the BSD Technical handbook for non-domestic buildings Standard 6.5 requires that "every building must be designed and constructed in such a way that the artificial or display lighting installed is energy efficient and is capable of being controlled to achieve optimum energy efficiency of fixed lighting". The standard however does not apply to process and emergency lighting.

Emergency lighting standards

3

3.1 European standards

The European standards for emergency lighting are:

(a) BS EN 1838:2013 Lighting applications – Emergency lighting
(b) BS EN 50172:2004 (dual numbered and implemented as BS 5266-8:2004) Emergency escape lighting systems.

The United Kingdom, as a member of the European standards body CENELEC, is obliged to give these EN (Euro Norm) standards the status of a national standard without any alterations.

BS EN 1838 is a European standard that specifies common European standards for luminous requirements for emergency escape lighting and standby lighting systems installed in premises or locations where such systems are required. It is principally applicable to locations where the public or workers have access.

British Standard 5266-1 *Emergency lighting – Part 1: Code of practice for the emergency lighting of premises*, was revised in 2005 to remove any requirements that are within the scope of the two European standards. This new 2016 revision continues in that way.

Emergency lighting is also covered by Chapter 56 of BS 7671 "Safety services" which is based on IEC and CENELEC requirements, but the UK has a derogation to allow emergency lighting to comply with BS 5266 and BS EN 1838.

3.2 British Standards

The British Standards Institution standards series for emergency lighting is BS 5266:

3.2.1 BS 5266-1:2016 *Emergency lighting. Code of practice for the emergency lighting of premises*

This part of BS 5266 gives recommendations and guidance on the factors that need to be taken into account in the design, installation and wiring of electrical emergency lighting systems, in order to provide the lighting performance needed for safety of people in the building in the event of failure of the supply to the normal lighting.

This British Standard applies to emergency lighting systems used to:

(a) assist occupants to leave a building during an emergency;

(b) help protect occupants if they stay in a building during an emergency; and

(c) help occupants to continue normal operations in the event of failure of the supply to the normal lighting.

This part of BS 5266 also gives recommendations for lighting in areas with fixed seating. In addition, this part of BS 5266 is not applicable to dwellings; however, its provisions are applicable to common access routes within blocks of flats or maisonettes.

3.2.2   BS 5266-2:1998 *Emergency lighting. Code of practice for electrical low mounted way guidance systems for emergency use*

This part of the BS 5266 series, provides recommendations for the planning, design, installation and servicing of electrical low mounted way guidance systems, for use within emergency lighting systems. It is intended to cover the use of low mounted way guidance systems for use in premises where such use has been agreed by all interested parties including the enforcing authority.

3.2.3   BS 5266-4:1999 *Emergency lighting. Code of practice for design, installation, maintenance and use of optical fibre systems*

BS 5266-4:1999 gives recommendations and guidance on the design, installation, maintenance and use of optical fibre emergency lighting systems. It is applicable to optical fibre emergency lighting systems for escape route lighting, including open area lighting. It is also applicable to optical fibre systems used for standby lighting when the system is also used as part of the emergency escape route lighting.

Note: This Part of BS 5266 series is to be used in conjunction with BS 5266: Part 1 and BS 5266: Part 5.

3.2.4   BS 5266-5:1999 *Emergency lighting. Specification for component parts of optical fibre systems*

BS 5266-5:1999 specifies requirements for optical fibres, light guides, connectors, emission end mounting arrangements and light sources to be used in optical fibre emergency lighting systems. Constructional and performance requirements are given, including performance under fire conditions.

The standard is applicable to the component parts of an emergency lighting system using optical fibre lightguides to distribute light from a light source to one or more lighting positions remote to that light source.

The standard specifies the use of optical fibres with end illumination and end emission. It is not applicable to optical fibres with end illumination and side wall emission.

Note: This Part of BS 5266 series is to be used in conjunction with BS 5266: Part 1 and BS 5266: Part 4.

3.2.5   BS 5266-6:1999 *Emergency lighting. Code of practice for non-electrical low mounted way guidance systems for emergency use. Photoluminescent systems*

BS 5266-6:1999 gives recommendations for the planning, design, installation and servicing of photoluminescent low mounted way guidance systems, for use within emergency lighting systems.

3.2.6   BS EN 1838:2013 (BS 5266-7) *Lighting applications – Emergency lighting*

BS EN 1838 is a European Standard that specifies the luminous requirements that have been agreed in Europe for emergency escape lighting and standby lighting systems installed in premises or locations where such systems are required. It is principally applicable to locations where the public or workers have access. This standard gives the luminous requirements for emergency lighting systems and it is to be read in conjunction with BS 5266-1. BS 7671 makes reference to Safety Systems in Chapter 56 but the majority of requirements in the European Chapter 56 do not apply in the UK as we have a separate standard BS 5266.

Note: The previous 1999 edition of BS EN 1838 was also dual numbered BS 5266-7:1999 but the reference to BS 5266-7 no longer applies in the 2013 edition.

3.2.7   BS EN 50172:2004, BS 5266-8:2004 *Emergency escape lighting systems*

This dual numbered standard sets out the requirements for emergency lighting in buildings – such as office and multi-storey buildings that are open to the public. It looks at the illumination of escape routes and safety signs if the normal supply fails. These standards specify the minimum requirements of emergency lighting based on the size, type and usage of the relevant premises. They also apply to standby lighting used as emergency escape lighting and provide practical guidance to make it more effective in an emergency.

Product (equipment) standards include:

1. BS EN 60598-2-22:2014 *Luminaires. Particular requirements. Luminaires for emergency lighting*
2. BS EN 50171:2001 *Central power supply systems*

BS EN 50171:2001 specifies the general requirements for central power supply systems for an independent energy supply to essential safety equipment. This standard covers systems permanently connected to AC supply voltages not exceeding 1000 V and that use batteries as the alternative power source.

The central power supplies are intended to energize emergency escape lighting in the case of failure of the normal supply, and may be suitable for energizing other essential safety equipment.

3. BS EN 62034:2012 Automatic test system for battery powered emergency escape lighting.

3.3   Design information

The three core standards, Parts 1 and 8 of BS 5266 and BS EN 1838, need to be considered together, see Figure 3.1.

Part 1 is a code of practice making recommendations; Part 8 and BS EN 1838 are the European system standards. The system standards are supported by product standards, particularly those listed above and shown in Figure 3.1.

Note: Persons carrying out emergency lighting designs will need to obtain a copy of Parts 1 and 8 and BS EN 1838 as the requirements of the system standards (Part 8 and BS EN 1838) are not reproduced in the Code of Practice (Part 1).

▼ **Figure 3.1** Relationship between the core emergency lighting standards (Parts 1, 7 and 8 of BS 5266) and product standards

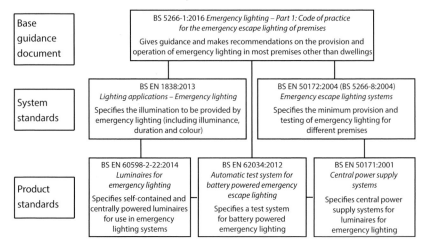

3.4   Definitions

The definitions in the standards are not completely identical. The definitions below are mostly those of BS EN 1838 and BS 5266-8:

3.4.1 Emergency lighting

Lighting provided for use when the supply to the normal lighting fails (Clause 3.6 of BS 5266-1). Emergency lighting includes emergency escape lighting, emergency safety lighting and standby lighting:

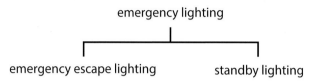

▼ **Figure 3.2**   Types of emergency lighting

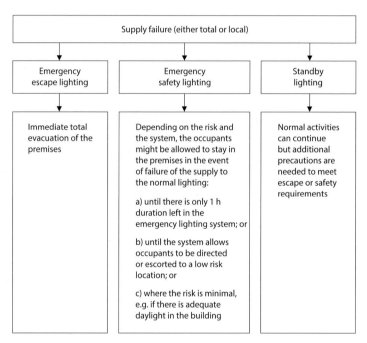

3.4.2 Standby lighting

That part of emergency lighting provided to enable normal activities to continue substantially unchanged (Clause 3.7 of BS EN 1838 and Clause 3.18 of BS 5266-1).

3.4.3 Emergency escape lighting

That part of emergency lighting that provides illumination for the safety of people leaving a location or attempting to terminate a potentially dangerous process before doing so (Clause 3.3 of BS EN 1838 and Clause 3.5 of BS 5266-1).

Emergency escape lighting includes escape route lighting, open area lighting and high risk task area lighting.

3.4.4 Emergency safety lighting

That part of emergency lighting that provides illumination for the safety of people staying in a premises when the supply to the normal lighting fails (Clause 3.7 of BS 5266-1).

If the premises are not fully evacuated in the event of a failure of the normal lighting supply additional measures to those used to provide emergency escape lighting would be required to provide adequate illumination to those remaining. This may include coverage of rooms that will be occupied during the supply failure or the need for higher light levels as determined by a risk assessment for the activities taking place. Guidance given in the standard for the use to conduct activities can provide a useful guide of suitable levels of illumination (Annex E).

3.4.5 Escape route

Route designated for escape to a place of safety in the event of an emergency. (Clause 3.8 of BS 5266-1).

(Local Building Regulations will make specific requirements for the construction of and finishes in an escape route).

In a new building, adequate escape routes and places of safety will be identified and specified by the design team to comply with the necessary Building Regulations. In an existing building, uses may change over time and any refurbishment designs must take the requirements for adequate escape routes and places of safety into account, and again this must comply with the necessary Building Regulations. During the life of a building (other than a dwelling) a regular fire risk assessment must be carried out by a competent person and this must determine whether the identified escape routes are still adequate or improvements made. (See BS 9999 for further guidance).

3.4.6 Emergency escape route lighting

That part of emergency escape lighting provided to ensure that the means of escape can be effectively identified and safely used at all times when the location is occupied (Clause 3.3 of BS 5266-8; this is called escape route lighting in Clause 3.4 of BS EN 1838).

The objective of escape route lighting is to enable safe exit for occupants by providing appropriate visual conditions and direction finding on escape routes and to ensure that fire-fighting and safety equipment can be readily located and used.

3.4.7 Open area lighting (anti-panic)

That part of emergency escape lighting provided to avoid panic and provide illumination allowing people to reach a place where an escape route can be identified (in some countries this is known as anti-panic lighting). (Clause 3.14 of BS 5266-1 and Clause 3.5 of BS EN 1838).

3.4.8   High risk task area lighting

That part of emergency escape lighting that provides illumination for the safety of people involved in a potentially dangerous process or situation and to enable proper shut down procedures for the safety of the operator and other occupants of the premises (Clause 3.10 of BS 5266-1 and Clause 3.6 of BS EN 1838).

3.5   BS 5266 Emergency lighting

In this summary of the requirements of the British Standard, the structure of BS 5266 Part 1 (Code of Practice for the emergency lighting of premises) has been followed. The requirements of Parts 7 and 8, that is, the requirements of the European standards, have been inserted as appropriate.

It should be noted that the scope of application of BS 5266-1:2016 includes theatres, cinemas and a wide range of other premises used for entertainment or recreation (see the Scope of the standard).

The standard specifies requirements for 'emergency lighting', which by definition is 'lighting provided for use when the supply to the normal lighting fails'. Emergency lighting includes escape lighting, safety lighting and standby lighting. These are discussed further below.

Emergency escape lighting is required not only on complete failure of the supply to the normal lighting but also on a localized failure if such a failure would present a hazard, for example a lighting circuit on a stairway.

(Clause 4.5 of
BS EN 1838)

3.5.1   Standby lighting

Standby lighting, that part of emergency lighting provided to enable normal activities to continue substantially unchanged, is within the scope of the standard, in particular Part 7 (BS EN 1838). However, few requirements are listed. When standby lighting is used as part of the emergency escape lighting, it must comply with the requirements for emergency escape lighting. Where standby lighting is provided at an illumination level lower than the minimum normal lighting level then it is to be used for shut down purposes only.

3.5.2   Emergency safety lighting

Although not a new concept this has not been included in the standard previously. If the lighting in a premises, or a part of a premises, fails it may be due to an electrical fault or a power cut and it is not 'automatically' an emergency where evacuation is necessary – especially in places such as a care home etc.

BS 5266-1:2016 now addresses the risks that occupants may face if they remain in the building while there is a failure of the normal lighting. It may be decided by a responsible person that they can stay for up to two hours whilst the fault is being rectified then to evacuate during the remaining one hour capacity if the fault cannot be repaired. In these cases the system must be correctly tested to ensure that the three hour design capacity will be available. (This procedure has been in common use for cinemas and theatres for a considerable time). If a building is not evacuated immediately, additional measures to those normally required might need to be included in the risk assessment. These may include the provision of emergency lighting in rooms where occupants are able to stay put, and also higher light levels in some general areas.

In some buildings, evacuation may be impracticable until an emergency occurs and in these cases it may be decided to conduct the occupants to a relatively safe location. This procedure still needs provision of protection in case an emergency requiring evacuation occurs during the supply failure. This can be provided by use of switched luminaires retaining one hour evacuation capacity or by trained fire wardens who are equipped with safety hand lamps complying to BS EN 60598-2-22.

In some locations during daylight the risk may be considered to be minimal in which case supervisors will need to ensure evacuation takes place before daylight fails, and these premises should be inspected to ensure that occupants escape routes do not pass through any parts of the premises without natural illumination by windows etc.

The introduction section of BS 5266-1 explains the use and formats of emergency safety lighting, and depending on the risk and the emergency lighting system the occupants might be allowed to stay in the building in the event of failure of the supply to the normal lighting if the risk is minimal e.g. if there is adequate daylight in the building or until there is only one hour duration left in the emergency lighting system. Procedures for maintaining safety should be determined by responsible persons, including actions to be taken as the end of emergency duration approaches, how to warn the occupants if they then need to evacuate the premises and how to direct or escort the occupants to escape routes or safe refuges, if these are to be used.

3.5.3   Emergency escape lighting

(Clause 4.1 of BS 5266-8)

Emergency escape lighting is that lighting provided in the event of failure or partial failure of the electricity supply so as to provide sufficient illumination for the safety of people leaving a location including making safe a potentially dangerous process.

Note: Not all routes out of a building will be designated escape routes.

Escape lighting is not designed to allow normal operations to continue in the building. It is designed to allow making safe a potentially dangerous process and safe exit from the building.

The standard requires that the installation shall ensure that emergency escape lighting fulfils the following functions:

(a) to illuminate escape route signs;
(b) to provide illumination onto and along such routes as to allow safe movement towards and through the exits provided to a place of safety, including open area lighting;
(c) to ensure that fire alarm call points and fire safety equipment provided along the escape routes can be readily located and used;
(d) to permit operations concerned with safety measures, including 'high risk task area lighting'; and
(e) to provide illumination to and at a place of safety.

Figure 3.3 summarises the functional requirements of emergency escape lighting.

▼ Figure 3.3   Functional requirements of emergency escape lighting

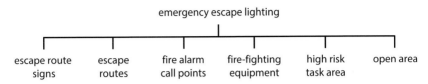

3.5.4   High risk task area lighting

High risk task area emergency lighting is provided for the safety of people involved in potentially high risk work or high risk situations so as to avoid hazards and to allow the proper shut down of work that might otherwise result in danger.

3.5.5   Open area (anti-panic) lighting

Open area lighting, sometimes called anti-panic lighting, is installed to reduce the likelihood of panic in relatively large areas (over 60 m^2) with perhaps undefined escape routes such as in halls. It should be considered for smaller areas, if the number of people in the area is likely to cause disorientation or panic.

It is now clarified in BS 5266-1 that open areas which have an inner room should be provided with emergency lighting as this area is an escape route from the inner room. Guidance is given that risk assessors should consider the need for emergency lighting to compensate for the increased risks resulting from high levels of occupancy, underground or windowless rooms or those with an escape route passing through or those which require switching off other equipment before leaving.

3.5.6   BS 5266-1 Borrowed light

'Borrowed light' is defined as light obtained from an adjacent reliable source that is expected to be available at all material times; for example, from a local emergency luminaire.

Design

4

(Clause 5 of
BS 5266-8)

4.1 Introduction

Emergency escape lighting is a part of the fire safety system (alarms, detectors, sprinklers, gas systems, etc.) required under UK fire safety legislation and is designed as an integral part of the system. This legislation imposes a duty on persons including employers and other persons who have control of premises to carry out risk assessments and to take precautions so as to ensure – as far as is reasonably practicable – the safety of persons working in, using or visiting the premises.

The strategy for the evacuation of the building must be decided before designs for the fire detection and alarm system and the emergency lighting installation are started. Part B of the Building Regulations makes specific requirements for escape routes and other safety provisions as a part of the construction of a building. These routes must be agreed with and approved by the Local Authority Building Control or other authorized inspectors or verifiers as a part of the building construction approvals process. Designers must know the designated escape routes to be taken to places of safety to ensure that they are appropriately indicated and illuminated. The users (employer) will also need to understand the strategy – so that they can ensure that the escape routes are properly maintained and kept clear and an escape will proceed accordingly to the plan.

As noted previously, it is required to carefully consider the layout of escape routes, safety equipment and emergency illumination in premises to ensure that adequate provision is made for safe escape from a building to a place of safety. During the life of a building there may well be changes of use of the whole or parts of the building (a storage area converted into office space for example) and as a part of this change it is necessary for the dutyholder or responsible persons in charge of the premises to ensure that adequate provision continues to be made for safe escape from a building to a place of safety. Escape routes must be adequate to allow safe escape as required by the local Building Regulations, and new escape routes may be required – with adequate emergency lighting. Any changes should be identified and considered as a part of the premises regular fire risk assessments.

Just because a person is out of a building it does not mean they are safe, and the Building Regulations and BS 9999 require escape "to a place of safety" where the persons escaping can be sure they are no longer at risk from fire or dangers brought about by the effects of fire (for example, dangerous structures). Although in some large buildings (for example, hospitals) "horizontal evacuation" may allow escape into an adjacent safe part of the building complex generally escape is to outside the building. Emergency illumination and escape signage will then be required on escape routes outside the building to guide and illuminate the way to a nominated assembly area "place of safety" where persons are checked to ensure that all are accounted for. It may well be that local street lighting can provide some or all of this external illumination, but these provisions should be regularly reassessed to confirm whether the street lights will still be illuminated at all times the premises is in use; if they will not, alternative emergency illumination arrangements should be made. Any requirements should be identified and considered as a part of the premises regular fire risk assessments.

The selection of escape routes is an integral part of the building design requiring, as well as lighting and signs, other provisions for escape to a place of safety (such as suitable exits, panic bolts, etc.).

Emergency escape lighting is required to provide:

(a) escape route lighting including illumination of:
  (i) escape route signs;
  (ii) escape routes to allow safe movement towards and through the exits provided to a place of safety, including 'open areas' for anti-panic lighting; and
  (iii) fire alarm call points and fire safety equipment provided along escape routes so that they can be located and used;
(b) illumination of high risk task areas (see paragraph 3.5.4) in the event of loss of lighting to allow danger to be avoided and shut down procedures to be implemented as necessary (to facilitate operations concerned with safety measures).

The objective of escape route lighting is to:

(a) facilitate safe escape for occupants by illumination and direction finding of escape routes as necessary; and
(b) enable fire-fighting and safety equipment to be readily located and used.

(Clause
4.1 of BS
5266-8)
(Intro to
BS 5266-1)

4.1.1   Risk assessment

The Management of Health and Safety at Work Regulations 1999 require every employer to carry out a risk assessment of his/her premises (see Section 1.4). The measures necessary to provide for the safety of staff and other persons using the premises will include the provision of safe means of escape, and are likely to require fire detection and alarm, provision for fire-fighting, and include emergency lighting. The requirements of legislation (Chapter 1) and the Building Regulations in particular (Chapter 2) must be considered when relevant. The risk assessment will identify the 'high risk task areas'.

Risk assessment guidance is given on the HSE site:

www.hse.gov.uk/toolbox/fire.htm

Fire safety guidance is given on the UK Government website at:

https://www.gov.uk/workplace-fire-safety-your-responsibilities/who-is-responsible

4.1.2 Failure of normal supply to part of premises

(Clause 5.2 of BS 5266-8)

Local emergency lighting is required to operate in the event of failure of any part of the premises normal lighting system.

Non-maintained and combined non-maintained emergency luminaires are required to operate in the event of failure of a normal lighting final circuit. In all situations the local emergency escape lighting must operate in the event of failure of normal supply to the corresponding local area. (For centrally supplied systems this will require a control system with sensors and contactors for each luminaire or circuit.) (See Section 4.8 for emergency luminaire definitions).

If the normal lighting circuits are interleaved such that in a large room or corridor the luminaires are on different circuits, the risk of a complete loss of lighting is reduced.

4.1.3 Response times

Specifically, BS EN 1838 requires:

(Clause 4.2.6)
(a) emergency escape lighting to reach 50 % of the required illuminance level within 5 s and full required illuminance within 60 s;

(Clause 4.3.6)
(b) open area (anti-panic) lighting to reach 50% of the required illuminance within 5 s and full required illuminance within 60 s;

(Clause 4.4.6)
(c) high risk task area lighting to provide the full required illuminance permanently or within 0.5 s depending upon application.

There is no response time listed in BS EN 1838 for emergency safety lighting but it should have the same performance as both emergency escape and open area lighting. This should be agreed with and approved by the Local Authority Building Control or other authorized inspectors or verifiers and the performance confirmed on site by measurement during commissioning.

If standby generators are used to supply the emergency lighting, they must comply with these requirements and supply the emergency lighting load automatically within 5 s for normal applications or 0.5 s for high risk areas (where this is acceptable). Usually it is not possible (or economic) to ensure that larger generators will start and take up the relatively small emergency lighting load within such times so a combination of battery back-up and generators is sometimes installed. (See Clause 7.3)

4

4.1.4 Duration

Minimum duration

The minimum duration for emergency escape lighting is:

(Clauses 4.2.5, 4.3.5 and 4.4.5 of BS EN 1838 and 6.7.3 of BS 5266-1)

(a) escape routes – 1 h;
(b) open areas (anti-panic) – 1 h; and
(c) high risk task areas – time to make safe and escape. – to be agreed with user and approved by the Local Authority Building Control or other authorized inspectors or verifiers at design and the performance confirmed on site by measurement during commissioning.

Requirement

One hour should be sufficient time to evacuate the largest and most complex buildings and reach a place of safety. If it were not, the escape strategy would need to be confirmed as the most effective. However, the duration must be sufficient for orderly escape as per the strategy plus allowance for any disruption to orderly evacuation likely to occur in an emergency.

Allowance should be made for some of the escape routes to be cut off and injured persons given attention.

Following the risk assessment, particularly for larger premises, emergency lighting that will remain in operation after the evacuation of the building has been completed may be necessary to enable searches of the premises to be carried out and to allow return to the building after the emergency.

Further guidance is given in BS 5266-1, Clause 6.7.3 and at Annex E. A minimum duration of 3 h is required if premises are not expected to be evacuated immediately in the event of a supply failure, such as sleeping accommodation, places of entertainment or "emergency safety lighting" systems installations and in places that are expected to be reoccupied again as soon as the supply is restored without waiting for the system batteries to be recharged.

BS 5266-1 states, "A minimum duration of 1 h should be used only if the premises will be evacuated immediately on supply failure and not reoccupied until full capacity has been restored to the batteries".

4.1.5 Plans and records

(Clauses 5.1 and 6 of BS 5266-8 and 4 of BS 5266-1)

Plans showing the layout of the building and all

(a) escape routes including stairs, changes of direction or level, fire escape doors etc.;

(b) routes out of the building to the designated place or places of safety;

(c) the location of designated place or places of safety;

(d) fire alarm call points;

(e) fire-fighting equipment;

(f) safety and first-aid equipment;

(g) escape equipment (for example, evacuation chairs, fire telephones)

(h) escape signs;

(i) structural features that may obstruct the escape route;

(j) open areas for anti-panic lighting;

(k) high risk task areas;

(l) standby lighting requirements; and

(m) emergency safety lighting requirements

need to be prepared before an emergency lighting installation design can begin.

Upon completion of the installation, as-installed drawings of the emergency escape lighting installation are required to be prepared and a copy retained on the premises. The drawings need to identify all luminaires, and all the main components of the installation. There is a general requirement in BS 5266-8 that the drawings be updated as necessary (following any alteration to the installation) and signed by a competent person.

Plans and designs will need to be approved by Local Authority Building Control or other authorized inspectors or verifiers.

4.1.6 Emergency lighting design procedure

(Clause 10 of BS 5266-1)

The following procedure may be followed to determine the emergency lighting requirements:

(Introduction and para 4.1 of BS 5266-1)

(a) carry out a risk assessment, to identify the hazards that will require emergency lighting, including high risk task areas;

(b) refer to the evacuation strategy prepared for the fire detection and alarm system;

(Clause 4.1 of BS EN 1838)

(c) position signs and luminaires at primary escape locations with direction signs if necessary, see Section 4.2;

(Clause 4.1.2 of BS EN 1838)

(d) position luminaires to illuminate all points of emphasis and at additional locations;

(Clause 4.2 of BS EN 1838) (Clause 4.3 of BS EN 1838) (Clause 4.4 of BS EN 1838)

(e) add luminaires as necessary to illuminate the escape routes;

(f) add luminaires as necessary to illuminate the open areas;

(g) illuminate high risk task areas, and

(h) position safety signs (see Figure 4.1).

▼ **Figure 4.1** Design sequence for emergency lighting luminaires

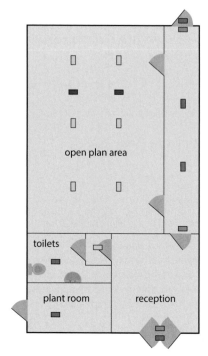

Design sequence, see 4.1.6

■ c) Position primary emergency escape route signs and luminaires at essential locations (4.2)

■ d) Position additional emergency escape lighting luminaires (4.3)

■ e) Add escape route luminaires as necessary to achieve minimum route illumination levels (4.4)

▢ f) Add open area lighting using spacing table (4.5)

■ g) Position high risk task area lighting (4.6)

A decision will have to be made whether centrally supplied luminaires or self-contained luminaires are to be installed. Self-contained luminaires may be more cost effective to install than centrally supplied, but maintenance costs may be higher; a payback period will need to be given to the client. Smaller installations, say 70 luminaires or fewer, tend to have self-contained luminaires.

(Clause 4.1 of BS EN 1838)

4.2 Primary escape route signs and luminaires

Where the direct sight of an emergency exit is not possible (for example, where there is a bend or fork in a corridor) an illuminated directional sign (or series of signs) should be provided to assist progression towards the emergency exit.

4.2.1 Signs

Chapter 7 provides guidance on safety signs as required by the Health and Safety (Safety Signs and Signals) Regulations.

4.2.2 Locations

Generally an escape lighting luminaire complying with BS EN 60598-2-22 is required near (outside) each exit door, and at positions where it is necessary to emphasize potential danger or safety equipment. The positions include the following:

Note: Clause 4.1 of BS EN 1838: For the purposes of this clause, 'near' is normally considered to be within 2 m measured horizontally.

a at each exit door intended to be used in an emergency

b* near stairs so that each flight of stairs receives direct light

c* near any other change in level

d* mandatory emergency exits and safety signs

e* at each change of direction

f at each intersection of corridors

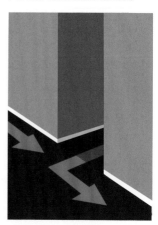

* **Note**: In locations b, c, d and e, whilst emergency lighting is required a direction sign may not be required.

g (no illustration): near each final exit and outside the building to a place of safety.

h near each first-aid post

i near each piece of fire-fighting equipment and call point

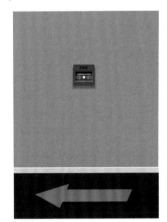

j (no illustration):   near escape equipment provided for the disabled.

k (no illustration):   near disabled person refuges and disabled person toilet alarm positions.

Where there is no direct sight of an exit and doubt may exist as to its location, a directional sign (or series of signs) shall be provided, placed such that a person moving towards it will be directed towards an emergency exit. There is no specific guidance as to how direction arrows for 'straight on' should point, but HSE document L64 (Safety signs and signals) Industry Committee on Emergency Lighting (ICEL) publications and BS 5499-4 all give guidance. Eventually it must be a 'common sense' choice on the installation taking into account the layout and the possibility of any misunderstandings, and a consistent approach should be taken through an installation.

An exit or directional sign shall be in view at all points along the escape route. Escape route signs should be selected and located in accordance with the guidance given in BS 5499-4:2013.

All signs marking exits and escape routes shall be uniform in colour and format, and their luminance, colormetric and photometric characteristics should comply with the guidance given in BS 5266-1 and BS 5266-7. Such characteristics are outside the scope of this guide and manufacturers' advice and guidance should be sought as required.

Maintained exit signs should be considered for applications where occupants may be unfamiliar with the building (see 4.8.3).

Escape routes should be unobstructed (see Approved Document M: Access to and use of buildings or regional equivalents); where this is not possible obstructions will need to be guarded and illuminated.

4

(Clauses
4.1 and
5.5 of
BS EN 1838
and
Clause
5.2.9.1 of
BS5266-1)

4.2.3 Viewing distances

Exit signs must be in the direct line of sight of persons in the building and within the viewing distance; see Section 4.9 for further information on dimensions etc. If they are not, an illuminated direction sign or signs must be installed to direct persons towards the exit.

Detailed guidance on viewing distances and sign size are outside the scope of this guide and manufacturers advice and guidance should be sought as required.

4.3   Additional emergency lighting luminaires

BS 5266-1 advises that emergency lighting should be provided for:

a   External areas in the immediate vicinity of exits to assist dispersal away from the exits to a place of safety, the illumination level being as for escape routes.

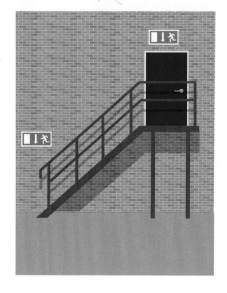

b Lift cars. These are not usually used for the escape route, but persons may be trapped in them in the event of a supply failure. Being confined in a small space in the dark without escape is not only unpleasant but may cause harm to those who are nervous or suffer from claustrophobia. Emergency illumination in the cars is recommended in evacuation lift cars as specified in BS EN 81-20 and it may also be prudent to install it in other passenger lift cars.

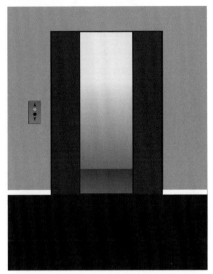

Note: If disabled people are given access to a building, one of their means of escape in emergency conditions may involve use of a lift or a place of safety or stairwell. Emergency lighting as specified for open area (anti-panic) lighting (see Section 4.5) is required in lifts in which persons may travel. The emergency lighting can be either self-contained or powered from a central supply, in which case a fire-protected supply will be required.

c   Moving stairways and walkways. These should be illuminated as if they were part of an escape route.

d   Toilet facilities. Facilities exceeding 8 m^2 gross floor area should be provided with emergency lighting as if they were open areas (see Section 4.5).

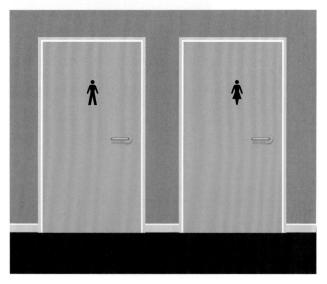

Toilets for disabled people and any multiple-closet facilities without natural or borrowed light available from an adjacent reliable source available at all material times should have emergency illumination. 'Borrowed' daylight would not be adequate if the facility could be in use at night.

Emergency lighting is not mandated in toilets designed for a single able-bodied person or en-suite toilets or bathrooms in hotel bedrooms.

e   Motor generator, control, plant and switch rooms. Battery-powered emergency lighting should be provided in all motor generator rooms, control rooms, plant rooms, switch rooms and adjacent to main control equipment associated with the provision of normal and emergency lighting to the premises.

f   (no illustration): Covered car parks. The pedestrian escape routes from covered and multi-storey car parks should be provided with emergency lighting.

4

4.4　Escape route illumination

4.4.1　Illumination levels

(Clause 5 of BS 5266-1, Clause 4.1.2, 4.2.1, 4.3.1 and 4.4.1 of BS EN 1838)

After positioning primary and additional signs, including supplementary direction signs, the illumination along the escape routes must be such that the means of escape can be identified and safely used in an emergency at any time when the location may be in use.

For routes that are permanently unobstructed and up to 2 m wide, the horizontal illuminance at floor level on the centre line of the escape route should not be less than 1 lx.

First-aid posts, fire-fighting equipment and call points are required to be illuminated to 5 lx vertical illuminance. Whilst informative only, Annex E of BS 5266-1 provides various levels of illuminance for locations including, but not limited to, kitchens, first aid rooms, refuge areas, and plant rooms).

BS EN 1838 specifies requirements for uniformity of illuminance and the disability glare limits, see below.

Note:　For design purposes it may not be reasonable to presume escape routes to be permanently unobstructed.

▼　**Figure 4.2** Uniformity of illuminance

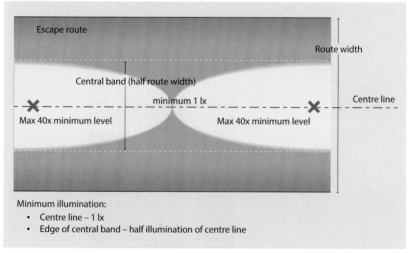

Escape route

Route width

Central band (half route width)

minimum 1 lx

Centre line

Max 40x minimum level　　Max 40x minimum level

Minimum illumination:
- Centre line – 1 lx
- Edge of central band – half illumination of centre line

For escape routes up to 2 m wide, the horizontal illuminances on the floor along the centre line of an escape route shall be not less than 1 lx and the central band consisting of not less than half of the width of the route shall be illuminated to a minimum of 50% of that value.

Along the centre line of the escape route the ratio of the maximum to the minimum illuminance shall be not greater than 40:1. The uniformity of illumination is expressed in terms of the ratio of the minimum illuminance to the average illuminance.

These illuminance levels are normally achieved by using the manufacturer's spacing tables to determine the maximum spacing between luminaires, see Table 4.1.

▼ **Table 4.1** Manufacturer's spacing table (typical)

Mode	Mounting height (m)	Lux level directly under	Luminaire spacing (m)			
Escape optic (Asymmetric)			Escape route 2 m wide 1 lux min			
Self-contained	2.5	2.7	–	–	17.1	7.8
	2.8	2.2	–	–	18.6	8.4
	3.0	1.9	–	–	19.8	8.8
Open area optic (Symmetric)			Open (anti-panic) area 0.5 lux min			
Self-contained	2.5	1.7	5.3	10.5	10.5	5.3
	2.8	1.4	5.7	11.5	11.5	5.7
	3.0	1.2	5.8	12.2	12.2	5.8
	4.0	0.67	4.9	12.5	12.5	4.9
Open area optic (Symmetric)			Open area 1 lux min			
Self-contained	2.5	1.7	4.3	9.4	9.4	4.3
	2.8	1.4	3.3	9.3	9.3	3.3
	3.0	1.2	3.2	9.2	9.2	3.2

Use of authenticated spacing tables

Where the transverse (long edge) to axial (short edge) spacing is needed, add one half of the transverse to transverse value to one half of the axial to axial value.

▼ **Table 4.2** Disability glare limits (Table 1 of BS EN 1838)

Mounting height above floor level, h (m)	Escape route and open area (anti-panic) lighting maximum luminous intensity, I_{max} (cd)	High risk task area lighting maximum luminous intensity, I_{max} (cd)
$h < 2.5$	500	1000
$2.5 \leq h < 3.0$	900	1800
$3.0 \leq h < 3.5$	1600	3200
$3.5 \leq h < 4.0$	2500	5000
$4.0 \leq h < 4.5$	3500	7000
$h \geq 4.5$	5000	10000

4.4.2 Existing installations

Some existing installations might have been designed and installed to an older standard which previously specified a minimum of 0.2 lx along the centre line of an escape route. These installations need to be reviewed to confirm that the illuminance provided continues to be acceptable for the current application.

(Clause 5.3 of BS 5266-8)

4.4.3 Compartment lighting

The illumination by the emergency escape lighting system of a 'compartment' of the escape route, see Figure 4.3, shall be from two or more luminaires so that the failure of one luminaire does not make the directional indication of the system ineffective or result in total darkness for part of the route. For the same reason, two or more luminaires shall be used in each open area with anti-panic lighting.

▼ **Figure 4.3** Compartment lighting

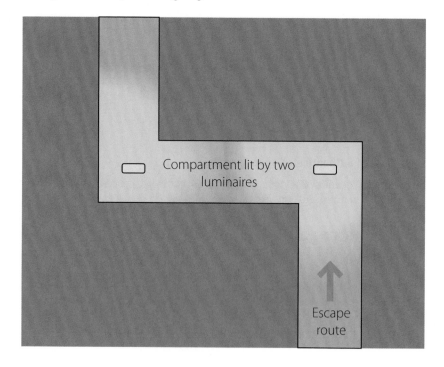

(Clauses 4.3 of BS EN 1838 and 4.4, 5.3 of BS 5266-8)

4.5 Open area (anti-panic) lighting

The objective of open area (anti-panic) lighting is to reduce the likelihood of panic and to enable safe movement of occupants towards escape routes by providing lighting and direction indication. It is used in open areas where escape routes are ill-defined, such as halls or premises larger than 60 m² floor area, or for rooms within rooms where the outer room becomes a part of the escape route from the inner room. It may be required in smaller areas if there is additional hazard such as use by a large number of people who might obstruct signs.

Where necessary, the illumination of the open area shall be from two or more luminaires so that the failure of one luminaire does not result in total darkness or make the direction indication of the system ineffective. For smaller areas, emergency illumination may not be necessary unless identified through a risk assessment.

The detailed requirements listed in Clause 4.3 of BS EN 1838 are:

(a) The horizontal illuminance (density of luminosity on a flat plane) shall be not less than 0.5 lx at the floor level of the empty core area which excludes a border of 0.5 m of the perimeter of thearea.

(b) The ratio of the maximum to the minimum anti-panic area lighting illuminance shall not be greater than 40:1.

(c) Disability glare shall be kept low by limiting the luminous intensity of the luminaires within the field of view. These shall not exceed the values in Table 1 of BS EN 1838 (Table 4.2 in this Guide) within the zone 60° to 90° from the downward vertical at all angles of azimuth (see Figure 2 of BS EN 1838).

(d) In order to identify safety colours, the minimum value for the colour rendering index R_a from a lamp shall be 40. The luminaire shall not substantially subtract from this.

(e) The minimum duration allowed for escape purposes shall be 1 h.

(f) The anti-panic area lighting shall reach 50 % of the required illuminance within 5 s and full required illuminance within 60 s.

(g) Open area lighting is required to toilets for disabled people.

(h) Open area lighting is required in a room with no direct access to an escape route.

Spacing tables again are usually used to determine the number and position of additional luminaires, see Table 4.3.

▼ **Table 4.3** Manufacturer's spacing table for open areas (typical)

Self-contained luminaires			Spacing, open area (m)			
Luminaire type	**Mounting in height (m)**	**Lux level directly under**	**0.5L** ⌐→□ ⊢—→⊢	**0.5L** □—↔□	**0.5L** □—△—□	**0.5L** □—↔⌐
Atlantic NM	2.5	1.66	2.5	9.1	5.4	1.4
NM	4.0	0.65	0.7	9.6	5.8	0.4
NM	6.0	0.29	–	–	–	–
M	6.0	0.29	2.4	8.8	5.0	0.4
M	4.0	0.55	0.9	8.0	5.0	0.6
M	6.0	0.25	–	–	–	–

[Courtesy of Menvier]

4.6 High risk task area lighting

(Clause 4.4 of BS EN 1838)

The objective of high risk task area lighting is to provide for the safety of people where there is a potentially dangerous process or situation such as to enable proper shut down procedures to be followed.

The detailed requirements of BS EN 1838 are:

(a) In areas of high risk the maintained illuminance on the reference plane shall be not less than 10 % of the required maintained illuminance for that task, however it shall be not less than 15 lx. It shall be free of harmful stroboscopic effects.

(b) The uniformity* of the high risk task area lighting illuminance shall be not less than 0.1.

(c) Disability glare shall be kept low by limiting the luminous intensity of the luminaires within the field of view. These shall not exceed the values in Table 1 (Table 4.2 in this Guide) within the zone 60° to 90° from the downward vertical at all angles of azimuth.

(d) In order to identify safety colours, the minimum value for the colour rendering index R_a of a lamp shall be 40. The luminaire shall not substantially subtract from this.

(e) The minimum duration shall be the period for which the risk exists, as agreed with the dutyholder or responsible person.

(f) High risk task area lighting shall provide the full required illuminance permanently or within 0.5 s depending upon application.

The illumination required is normally achieved by installing a conversion of the luminaire to provide emergency lighting or by installing a separate luminaire unit.

The 10 % required maintained illuminance for the task will normally be the limiting factor unless normal illumination levels are very low. Consequently, all fittings in the risk area need a standby ballast factor[†] of 10 % or every other one of 20 % etc. Ballast factor is a measure of the actual lumen output for a specific lamp-ballast system relative to the rated lumen output.

(Clause 4.5 of BS EN 1838)

4.7   Standby lighting

Standby lighting is that part of emergency lighting provided to enable normal activities to continue. When it is used for emergency escape lighting purposes it must comply with the requirements for escape lighting as required in BS 5266-1.

Where a standby lighting level lower than the minimum normal lighting is employed, the lighting is to be used only to shut down or terminate processes.

(Annex C of BS 5266-1)

4.8   Classification of emergency lighting systems

4.8.1   General

Emergency lighting systems are classified as given in BS EN 60598-2-22: 2014 by:

- **(a)** type of system;
- **(b)** mode of operation;
- **(c)** facilities; and
- **(d)** duration of emergency mode.

Table 4.4 lists the detailed classifications under these four headings.

▼ **Table 4.4** Emergency lighting classifications

Type		Mode of operation		Facilities		Duration of emergency mode for a self-contained system (minutes)
X self-contained	0	non-maintained	A	including test device		10
Z central supply	1	maintained	B	including remote test mode		60
	2	combined non-maintained	C	including inhibiting mode		120
	3	combined maintained	D	high risk task area luminaire		180
	4	compound non-maintained	E	with non-replaceable lamp(s) and/or battery		
	5	compound maintained	F	automatic test gear conforming to BS EN 61347-2-7 denoted EL-T		
	6	satellite	G	internally illuminated safety sign		

* The uniformity of illumination is expressed in terms of the ratio of the minimum illuminance to the average illuminance.

[†] Ballast factor is a measure of the actual lumen output for a specific lamp-ballast system relative to the rated lumen output.

Until quite recently, emergency lighting systems were categorized by the prefix 'M' for maintained and 'NM' for non-maintained systems, followed by a '/' and the number of hours duration claimed for the installation, for example for self-contained systems:

M/1 was a maintained 1 h duration system; this is now:

X	1	* * * *	60

NM/3 was a non-maintained 3 h duration system; this is now:

X	0	* * * *	180

* * * * in the third box stands for the facilities, details of which are added, as applicable, from Table 4.4 at the time of installation.

4.8.2 Types of luminaire

X Self-contained emergency luminaire

Luminaire providing maintained or non-maintained emergency lighting in which all the elements, such as the battery, the lamp, the control unit and the test and monitoring facilities, where provided, are all contained within the luminaire or adjacent to it.

Z Centrally supplied emergency luminaire

Luminaire for maintained or non-maintained emergency lighting which is energized from a central emergency power system that is not contained within the luminaire.

4.8.3 Mode of operation

Non-maintained emergency luminaire

Luminaire in which the emergency lighting lamps are in operation only when the supply to the normal lighting fails.

Maintained emergency luminaire

Luminaire in which the emergency lighting lamps are energized at all times when normal or emergency lighting is required.

Combined emergency luminaire

Luminaire containing two or more lamps, at least one of which is energized from the emergency lighting supply and the other(s) from the normal lighting supply. A combined emergency luminaire is either maintained or non-maintained.

Sustained luminaire

This description is not found in BS 5266-1 or -8 or BS EN 1838, but is used to describe a combined emergency luminaire with two or more lamps where at least one lamp operates in non-maintained mode and is only illuminated when the normal supply fails. The other lamp operates on the normal supply only. This is identical in functionality to having a Non-Maintained Luminaire and a Normal Luminaire both in the same housing.

Compound self-contained emergency luminaire

Luminaire providing maintained or non-maintained emergency lighting and also providing emergency supply for operating a satellite luminaire.

4.8.4 Facilities

Remote rest mode

Feature of a self-contained emergency luminaire that can be intentionally extinguished by a remote device when the normal supply has failed and that, in the event of restoration of the normal supply, automatically reverts to normal mode.

Inhibiting mode

Feature of a self-contained emergency luminaire that can be set independently from the condition of the normal power and therefore, when the building is unoccupied, a supply failure will not cause unwanted discharge.

4.8.5 Duration

(Clause 9 of BS 5266-1)

The classification (type, mode and duration) required for the system is agreed following the risk assessment and following consultation with the regulatory bodies. Guidance on minimum duration is given in BS 5266-1 and summarised in Table 4.5. The overriding consideration for the duration is that it is sufficient for the escape strategy.

▼ **Table 4.5** Emergency escape lighting duration

Class	Examples	Duration
Premises used as sleeping accommodation	This class includes such premises as hospitals, care homes, hotels, guest houses, certain clubs, colleges and boarding schools	3 h (note 1)
Non-residential premises used for treatment or care	This class includes such premises as special schools, clinics, dental practices and similar premises	3 h
Non-residential premises used for recreation	This class includes such premises as theatres, cinemas, concert halls, exhibition halls, sports halls, public houses and restaurants	3 h (note 2)
Non-residential premises used for teaching, training and research, and offices	This class includes such premises as schools, colleges, technical institutes and laboratories	1 h
	If any part of a building is used outside of normal weekday office hours	3 h
Non-residential public premises	This class includes such premises as town halls, libraries, shops, shopping malls, art galleries and museums	3 h
Industrial premises used for manufacture, processing or storage of products	This class includes such premises as factories, workshops, warehouses and similar establishments	1 h
Multiple use of premises		As for the most stringent class
Common access routes within blocks of flats or maisonettes		3 h
Covered car parks		3 h
Sports stadia	for spectators	3 h

Notes to Table 4.5:

1 Guidance for hospitals is given in the Department of Health *Electrical services* series of publications (see [31], [32] and [33] in the Bibliography to BS 5266-1). Advice on the application of this guidance in relation to that given in this standard should be obtained from the enforcing authority.

2 Where the normal lighting might be dimmed or turned off, a maintained or combined emergency escape lighting system should be installed. However, it is not necessary for the full emergency lighting level to be provided when the normal lighting system is functioning.

4.9 Requirements for safety signs

Minimum luminance

BS EN 1838, BS 5499-4:2013 and BS EN ISO 7010:2012+A5 specify requirements for:

(a) colour;
(b) luminance;
(c) ratios of colours; and
(d) ratios of luminance.

These are best achieved by using luminaires complying with BS EN 60598-2-22.

▼ **Figure 4.4** Escape route sign (see also Figure 7.6)

The measurement of luminance and luminance ratios is not something that is easily done on site and should preferably be assured by the manufacturer.

Maximum viewing distances are:

(a) internally illuminated signs 200 × height of pictogram (luminaire facia height);
(b) externally illuminated signs 100 × height of pictogram (luminaire facia height); and
(c) for example, an internally illuminated exit sign with pictogram height of 20 cm has a maximum viewing distance of 200 × 20 cm, that is 40 m (see Figure 4.5).

Detailed guidance on viewing distances and sign size are outside the scope of this Guide and manufacturers' advice and guidance should be sought as required. Where the angle of viewing of the escape route sign is not predominantly normal to the sign, the factor of distance reduced by a multiplying factor of cosine of the angle of viewing should be used.

▼ **Figure 4.5**   Escape route sign viewing distance

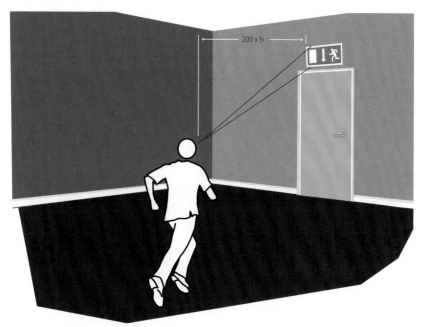

4.10   Automatic test systems

4.10.1 Introduction

Automatic test systems are available for emergency lighting luminaires. Luminaires with a diagnostic interface unit can be specified. These can allow automatic testing on an individual luminaire basis or automatic testing as initiated by a central control unit. Such test systems can be cost-effective as maintenance regimes are otherwise labour intensive.

(See also Section 6.11 for standard types and user requirements.)

BS 5266-1 advises that in premises where occupants are expected to stay in place through a supply failure, an automatic test system is used. Also if the responsible person is not able to ensure that regular emergency lighting testing is carried out manually BS 5266-1 advises the use of an automatic system to perform the test at the required intervals. In any event, in should be appreciated that automatic test systems are not "fit and forget" and the output of a test sequence must be reviewed to ensure that all faults identified are attended to. The automatic system must be suitable for the installation and used correctly and maintenance staff must be trained in its operation and to understand the output.

4.10.2 Central control testing systems

All the luminaires are fitted with addressable interface modules and connected by a communication medium such as data bus cable, see Figure 4.6. They can control both self-contained and central power systems.

▼ **Figure 4.6**   Automatic test system

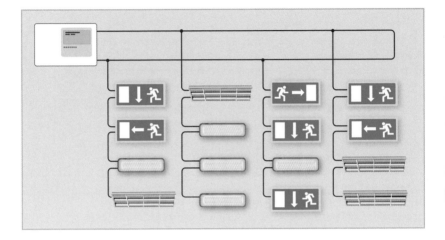

Note: The system may be fitted with data cable isolators, printer connections, PC interface etc.

Luminaires can be grouped for testing purposes so that, in the event of power failure after testing, a complete area is not left with run-down batteries, but the discharged luminaires are distributed around the building so maintaining effective signs and illumination.

See also Section 6.11.

4.10.3 Stand-alone luminaire testing

Luminaires fitted with a suitable testing module can be programmed to carry out tests on an individual luminaire basis giving an external signal if a fault is detected, or have testing initiated by a hand-held device.

Luminaires can be individually programmed to carry out maintenance checks by a hand-held programmer. This type of system is suitable for use with self-contained luminaires only.

Electrical installation

5

(Clauses 8 of BS 5266-1 and 4.1 of BS 5266-8)

5.1 Basic requirement

It is a requirement of BS 5266-8 that emergency escape lighting shall be activated not only on complete failure of the supply to the normal lighting but also on localized failure, such as a final circuit failure. Consequently, if the emergency escape lighting (that is non-maintained, self-contained or non-maintained from a central battery system) is supplied from a central supply there must be detection in final lighting circuits to detect loss of supply.

(Clause 8.1 of BS 5266-1)

5.2 Self-contained luminaires

Self-contained emergency luminaires do not require fire-protected cable supplies.

Self-contained luminaires have only one special requirement for their wiring other than that it must comply with BS 7671. Preferably, the wiring to the self-contained unit should not include or incorporate a plug and socket, unless precautions are taken to prevent accidental (or malicious) disconnection. There is no requirement to comply with the special requirements of BS 7671, Chapter 56 (safety services) for the wiring to self-contained luminaires.

Often it is recommended that the wiring to self-contained emergency luminaires be the same as that for the normal lighting because if fire-resistant cabling is used for the emergency luminaires but not for the normal luminaires, a fire may cause failure of supply to the normal luminaires but not to the emergency luminaires and then in unusual circumstances they may not switch on and in such circumstances there is a risk they may not operate when needed.

5.3   Central power supply systems

5.3.1   General

The wiring of central supply installations must comply with BS 7671. Chapter 56, safety services, would apply to the wiring for central supply systems and there are, additionally, particular recommendations in Clause 8 of BS 5266-1 for:

(a) fire protection of the cables (Clause 8.2.2);

(b) minimum cable size of 1.5 mm^2 (Clause 8.2.7);

(c) voltage drop (4 % maximum) (Clauses 8.2.7, 8.3.5 and BS 7671 or manufacturers guidance);

(d) wiring systems – conduit, trunking etc. metal or non-flame propagating) (Clause 8.2.5);

(e) segregation (Clause 8.2.6);

(f) joints (Clause 8.2.4);

(g) isolators, switches and protected devices (Clause 8.3.2);

(h) warning labels (Clause 8.3.4); and

(i) cable support, protection and fixing (Clauses 8.2.3, 8.2.6, 8.2.9, 8.2.11 and BS 7671 Part 5).

These recommendations are considered below.

5.3.2   Wiring systems

The Construction Products Regulations (CPR) requirements for cables came in to force on the 1st July 2017 and it is obligatory for cables, having an intended use for permanent installation in buildings and construction works, to be accompanied by a Declaration of Performance (DoP) for Reaction to Fire performance of the cables in accordance with stated CPR performance classifications, and to have CE marking under the CPR. All types of power, control and communication cables are covered and it is the responsibility of the cable manufacturer (the person who places the product on the market in the EU) to provide the DoP and to apply CE marking. (Cables having Resistance to Fire, meaning retention of functionality during a fire, are not covered by the requirement now being introduced. They are scheduled to be covered at an, as yet, unknown date in the future.)

The CPR covers the way in which the product is placed on the market. It does not say how and where a particular product should be used. Crucially it does not say what class of product should be used in any given circumstance. Unlike many other EU countries, the UK has no legislation that determines the level of performance required for the cables with respect to reaction to fire and it is for the client, designer and installer of the emergency lighting system to select and specify a suitable cable or cables for the installation, and it must be noted that the reaction to fire performance is only one of several factors that must be considered.

A second section of the statutory CPR requirements is "resistance to fire" and this has yet to be implemented. It is consequently understood that as all cables used in emergency lighting installations (except for self-contained luminaires as noted in Section 5.2) are required to be wired using fire-resistant cables they are not yet covered in the CPR requirements. There is, however, currently no proposed date for the implementation of the CPR "resistance to fire" requirements so designers and installers will need to keep themselves informed of any future legislation changes.

Cables

Clause 8.2.2 and 8.2.3 of BS 5266-1 covers cable selection requirements and cable systems and requires appropriate non-combustible methods of cable support. It specifies two classifications:

1. Emergency lighting cables with an inherently high resistance to attack by fire. The cables should, as a minimum, have a duration of survival of 60 minutes when tested in accordance with BS EN 50200:2015.
2. Enhanced emergency lighting cables with an inherently high resistance to attack by fire. The cables should, as a minimum, have a duration of survival of 120 minutes when tested in accordance with BS EN 50200:2015.

BS 5266-1, Clause 8.2.2 advises that emergency lighting cables with an inherently high resistance to attack by fire that may be suitable include:

(a) mineral insulated cables to BS EN 60702-1, with terminations complying with BS EN 60702-2;
(b) fire-resistant electric cables having low emission of smoke and corrosive gases to BS 7629; and
(c) armoured fire-resistant electric cables having thermosetting insulation and low emission of smoke and gases to BS 7846.

In addition to the 60 minutes survival duration the cables should meet the 30 minutes survival time when tested in accordance with BS EN 50200:2015, Annex E.

It is advised to seek confirmation from the cable supplier that these cables are not under the CPR requirements and the requirement for either standard or enhanced systems, as no cable standard specifically calls up this test.

Note: These cables meet the recommendations of Approved Document B, see Section 2.5.3 of Chapter 2, provided they follow a route selected to pass only through parts of the building in which the fire risk is 'negligible' and they 'should' be separate from any circuit provided for another purpose. Designers in Scotland, Wales and N Ireland should consult their local appropriate equivalent building regulations requirements.

BS 5266-1 requires cables or cable systems with an inherently high resistance to attack by fire for connecting centrally supplied emergency escape lighting to the standby supply. Such cables and cable systems should have duration of survival of 60 minutes for standard systems and 120 minutes for enhanced systems when tested in accordance with BS EN 50200:2015. Such cables and cable systems must be suitably fixed to prevent collapse or damage in heat and fire conditions. Metal cable supports, clips and fixings should be utilized where there is a danger of collapsed cabling impeding escape or firefighters search and rescue.

Clause 8.2.2 of BS 5266-1 advises that enhanced emergency lighting cables with an inherently high resistance to attack by fire that may be suitable include:

(a) mineral insulated cables to BS EN 60702-1, with terminations complying with BS EN 60702-2;

(b) fire-resistant electric cables having low emission of smoke and corrosive gases to BS 7629-1; and

(c) armoured fire-resistant electric cables having thermosetting insulation and low emission of smoke and gases to BS 7846.

It is advised to seek confirmation from the cable supplier that these cables are not under the CPR requirements and the requirement for either standard or enhanced systems, as no cable standard specifically calls up this test.

Cable systems

Clause 8.2.2(c) of BS 5266-1 states that cable systems with an inherently high resistance to attack by fire that may be suitable include:

"Fire-resistant cables enclosed in screwed steel conduit, such that the cable system has a duration of survival of 60 min when tested in accordance with IEC 60331-3."

Designers and installers must confirm with their supplier that the cables do meet the CPR requirements and the duration of survival time of 60 minutes.

Note 3 in Clause 8.2.2 of BS 5266-1 states that emergency escape lighting systems for certain large and complex buildings might require enhanced emergency lighting cables or cable systems capable of longer fire survival times, which might include unsprinklered buildings or parts of buildings.

Again designers and installers are advised to confirm with their supplier that the cables are not under the CPR requirements and the duration of survival time of 120 minutes.

Additional fire protection may be provided by the building structure if the cables are buried in the structure or installed where there is negligible fire risk and separated from any significant fire risk by a wall, partition or floor having at least a one hour fire resistance.

5.3.3 Cable cross-sectional area and voltage drop

(Clauses 8.2.7 and 8.3.5 of BS 5266-1 and BS 7671 Appendix 4)

Cable conductors are required to be of copper with a minimum nominal cross-sectional area of 1.5 mm^2.

Voltage drop between the central battery or generator and centrally supplied luminaires should not exceed 4 % of the system nominal voltage at the maximum rated current and at the highest working temperature if more detailed manufacturer's information is not available.

5.4   Cable support, fixings and joints

(Clause
8.2.3 of
BS 5266-1)

5.4.1   Cable support and fixings

Methods of cable support and fixings should be non-combustible, such that circuit integrity is not reduced below that afforded by the cable used. That is, they should be able to withstand a similar temperature and duration as the cable with the same water testing.

Chapter 52 of BS 7671 specifically requires all cables and wiring systems in a building to be securely fixed such that they will not be liable to premature collapse in heat/fire and impede the progress of persons out of the building or the access of firefighters for search and rescue. In effect, this recommendation precludes the use of only plastic supports and fixings where these would be the sole means of cable support and some metal supports and fixings should also be utilized where necessary to provide secure support and fixings (currently there is no specific guidance as to the use of plastic plugs for screw fixings).

Where support is provided by drop rods, the drop rod size should be calculated in accordance with BS 8519: *Code of Practice for selection and installation of fire resistant power and control cable systems for life-safety and fire-fighting applications*. At present wire rope suspension systems are not always tested to or certificated to this standard and as such designers have to make their own decisions if electing to use such devices.

(Clause 8.2.4
of BS 5266-1)

5.4.2   Cable joints

Cables should be installed without joints external to the equipment. Joints other than those within the system components such as the luminaires or control units should be constructed so that they have the same fire withstand (water and temperature) as the cable. Such a joint will need to be enclosed in a suitable box and labelled appropriately 'EMERGENCY LIGHTING', 'EMERGENCY ESCAPE LIGHTING' or 'STANDBY LIGHTING' and also with the warning 'MAY BE LIVE' (see also Regulation 537.1.2 of BS 7671 requiring a warning notice for any equipment supplied from more than one source. Emergency lighting circuits may be live when other circuits have been isolated).

Note: The standard sees joints as a last resort if suitable cabling is not available or a viable option. Also joints need to be porcelain or similar heat resisting construction and not traditional polythene or PVC insulated connectors of crimps.

(Clause
8.2.6 of
BS 5266-1)

5.5   Segregation

For central supply systems it is essential that the wiring to the emergency escape lighting is segregated from the wiring of other circuits, either by installation in a separate conduit, ducting or trunking or by a metal or other rigid, mechanically strong and continuous partition of full depth without perforation. This is primarily to prevent mechanical damage to such cables, when operatives may be working on other wiring systems in close proximity.

5.6 Continuity of supply to the emergency lighting

The installation designer must bear in mind that maintenance and repair of the electrical installation in general will require from time to time circuits of the electrical installation to be switched off. Switching off the supply to emergency lighting luminaires may switch on the standby supply and run down the batteries, and may as a consequence in the event of a fire put the occupants at risk. The installation needs to be arranged so that, as far as is practicable, maintenance to the electrical installation can be carried out without disconnecting the supply to the emergency lighting, or carrying out the work at times that will give the emergency lighting time to recharge before the building is reoccupied. This will generally mean that for an installation comprising self-contained units the lighting circuits may need to be supplied separately from power or other lighting circuits – see Figure 5.1.

The occupier may also wish to switch off the building supply for safety or energy saving. However, such switching must not switch off the supply to the emergency lighting (so as to run down the standby batteries).

▼ **Figure 5.1** Circuit arrangement for self-contained luminaires

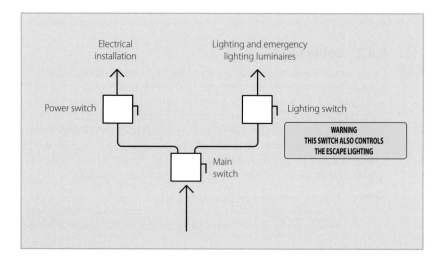

5.7 Isolation

For central supply systems, each isolating and switching device, etc. should be marked 'EMERGENCY LIGHTING' or 'STANDBY LIGHTING', as appropriate. For such systems it should be practicable to allow switching off of the general installation whilst maintaining the supply to the emergency lighting, in a similar way as it is necessary to maintain the supply to fire detection and alarm systems – see Figure 5.2.

▼ **Figure 5.2**   Circuit arrangement for central supply emergency lighting

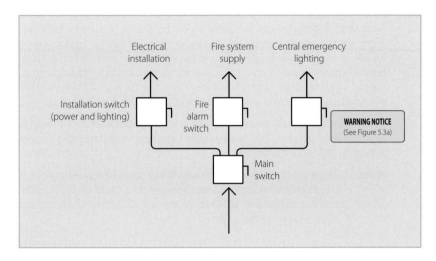

(Regulation
514.11 of
BS 7671
and Clause
8.3.4 of
BS 5266-1)

5.8   Warning notices

There is a general hazard associated with standby supply systems that switching off the circuits may not result in dead circuits as the standby supply may well make circuits live. Suitable warning notices must be installed. It must also be remembered that non-illumination of a luminaire or sign does not always indicate that the sign or circuit is dead. See Figure 5.3.

▼ **Figure 5.3**   Warning notices
   a

> **WARNING**
> **This equipment is also supplied from a standby supply.**
> **Switching off will not make all circuits dead.**

   b

> **WARNING**
> **Non-maintained luminaire.**
> **Live when lamp not illuminated.**

5.9   Test facility

All emergency luminaires require a suitable testing facility – whether manual or an automated test system and the facility should be able to be used for both the monthly short duration and the annual full duration testing. For manual testing, a key switch is usually provided for each lighting circuit, the operation of which interrupts only the live feed to either the self-contained luminaires on the circuit or provides only a "loss of supply" signal to central battery luminaires. Operating the key switch should not interrupt the normal lighting or to any other equipment that could cause a hazard.

A key switch is usually provided to allow security and to protect the system from unauthorized operation. The switch should be located in a convenient place such that testing can be easily carried out.

Note: Tripping a circuit-breaker, withdrawing a circuit fuse or providing a key switch at the distribution board to isolate the supply to a circuit are NOT usually suitable test methods as they will interrupt the normal lighting supply or the supply to other equipment that could cause a hazard.

5.10   Inspection and testing

Emergency lighting installations must be inspected and tested in accordance with the requirements of BS 7671 and an Electrical Installation Certificate issued together with schedules of inspection and schedule(s) of test results.

Inspection and testing of the emergency lighting installation as per BS 5266-1 (Clause 12) is also necessary, see Chapter 6.

5.11   BS 7671 Chapter 56 Requirements for Safety Services

Chapter 56 of BS 7671 has requirements for 'safety services'. These are generally as per BS 5266-1 where applicable to emergency lighting. The UK has a dispensation from the European regulations that BS 7671 is based on for Chapter 56 to allow emergency lighting to be installed to the requirements of another British Standard BS 5266.

Operation and maintenance

6

6.1 General requirements

Emergency lighting systems must be regularly inspected and tested and maintained to confirm that they are operational and that they are able to provide illumination for the designed time period – usually three hours. Also any ancillary equipment such as generators, battery chargers, etc. needs to be included in this regular inspection, testing and maintenance.

All luminaires should be functionally tested every month – or more frequently in areas with an arduous environment such as construction sites etc. Luminaires should also be cleaned regularly if in environments where dust can build up as this can reduce the light output and cause overheating of the luminaire.

A discharge test for the full rated duration of the installation should be performed on all luminaires at least annually. There is no requirement in BS 5266 for a six-monthly, 30-minute test; this will only reduce the life of batteries and certain types of lamps.

(Regulation 4(2) of EWR)

6.2 Legislation

The Electricity at Work Regulations 1989 (EWR) impose duties upon the employer, including:

> "As may be necessary to prevent danger, all systems shall be maintained so as to prevent, so far as is reasonably practicable, such danger."

The HSE's Electricity at Work Regulations 1989, Guidance on Regulations (HSR)25, advises regular inspection and maintenance and records of maintenance including test results. Inspection and, where necessary, testing of equipment is an essential part of any preventive maintenance programme. Practical experience of use may indicate an adjustment to the frequency at which any preventive maintenance needs to be carried out. This is a matter for the judgement of the dutyholder, who should seek all the information they need to make this judgement including reference to the equipment manufacturer's guidance.

6

6.3   Instructions

British Standard 5266-1 recommends that the designer includes that instructions on the operation and maintenance of the emergency lighting are prepared as a part of the design and installation schedule. It is intended that the instructions should be in the form of a manual, complete with installation layout drawings and schematic diagrams, to be retained by the occupier of the building. It was usual for such manuals to be paper based but now that digital storage and information systems are normal the manual can be computerized as long as the occupier/dutyholder and the maintenance staff can access and read the manual and drawings and amend any documents provided if modifications are made to the emergency lighting installation.

This manual is essential and must be obtained to form the starting point for the preparation of operation and maintenance procedures.

It is particularly important that operating instructions and manuals are provided for automatic testing systems.

6.4   As-installed drawings

On completion of the work, 'as-installed' drawings of the emergency escape lighting installation are required to be prepared for retention on the premises. The drawings are to identify all luminaires and the main components of the installation, and usually on larger installations with automated testing systems each emergency luminaire is given a unique identification. The drawings shall be signed by a competent person to verify that they are 'as-installed', and that the design meets the requirements of BS 5266-1. As discussed above the drawings can be computerized as long as the occupier/dutyholder and the maintenance staff can access and read the drawings and amend any documents provided if modifications are made to the emergency lighting installation.

Maintenance of the installation in accordance with BS 5266-1 requires that the drawings be updated as and when any changes or modifications are made to the installation.

The drawings are required to meet the requirements of Regulation 514.9.1 of BS 7671. It is to be noted that for simple installations, this regulation requires a durable copy of a schedule relating to each particular distribution board to be provided within or adjacent to each distribution board.

6.5   Handover

A competent person representing the designer and installer of the system on handover should:

> **(a)** train a person or persons representing the client on the necessary routine inspections and tests and if installed how to use the automatic test equipment features. This training may take some time but it is necessary for the maintenance staff to be properly trained and competent (it is unlikely that this could be achieved in a short demonstration at formal handover and the

maintenance staff should be involved during the installation commissioning stages before handover).

(b) provide necessary documentation for proper use of the system including design, installation and verification declarations, certification, logbook (see Clause 11 of BS 5266-1) and any other relevant information.

(c) explain the importance of keeping the logbook up to date and retaining the documents in an accessible safe place for reference as necessary.

(Clause 6 of BS 5266-8 and Clause 11 and Annex J of BS 5266-1)

6.6 Logbook

The standard requires a logbook to be retained on the premises in the care of a responsible person appointed by the occupier or owner. (The HSE's Electricity at Work Regulations 1989, Guidance on Regulations (HSR)25, imposes this responsibility on the dutyholder.)

The information required to set-out and be recorded in a logbook will vary from installation to installation. For example, the content provided for an installation that comprises a few self-contained luminaires is likely to be considerably less than that for a shopping mall or large industrial unit. In the former, a single page of installation-specific explanation and a manufacturer's data sheet might suffice, whereas, in the latter scenario, many volumes may be required.

So, before embarking on the production of a logbook, do consider the presentation format to be used, and whether it should be a stand-alone document or fully or partially included into a larger document. Another point for consideration is the media to be used: as in should it be presented in paper or electronic format; if it is to be electronic, is any particular software required.

Annex J of BS 5266-1 advises of information that should be recorded in the logbook, and the logbook should be used to record at least the following information:

1. Part 1: Contractual and legal details. This should be used to record the:
   (i) name and address of the installation;
   (ii) details of premises ownership (landlord, developer, tenant, third party, etc.) and extent of responsibility for the emergency lighting installation;
   (iii) details of leases, wayleaves, covenants, etc. affecting the premises and considered during the design stage;
   (iv) details of adjacent premises, common areas, surrounding public or private spaces, etc. which influenced the design;
   (v) details of local authority consents applicable to the premises (planning requirements or restrictions, building control sign-off, etc.); and
   (vi) details of who owns the copyright of the finished logbook.
2. Part 2: Project brief. This section should be used to include the details of the brief provided, setting out the reasons for commissioning a new emergency lighting installation or modifying an existing installation. Where applicable, the details of any specific requirements which affected the design or installation should also be recorded in this section.

3. Part 3: Risk analysis. This section should be used to provide a record of the risk analysis carried out and the specific risks which were addressed in the design. The detail should cover risks identified both inside and outside the premises, which might include:

   (i) details of toxic or noxious materials present in the premises

   (ii) hot surfaces

   (iii) processes which might be dangerous

   (iv) dangerous machinery, etc., which might have warning signs or safety instructions that need to be highlighted by emergency lighting.

4. Part 4: Equipment details. This section should be used to record details of the manufacturer, type, model/part number, serial number and similar information for all components forming the installation.

5. Part 5: Design and/or modification. This section should be used to set out details of the installation design or modification, including:

   (i) reasons for selecting particular items of equipment

   (ii) considerations relating to the choice of luminaires

   (iii)why the chosen wiring methods were selected

   (iv) other relevant information which might be required for an understanding of the system design and its future modification.

6. Part 6: Calculations. This section should be used to set out details of the calculations carried out and results obtained.

7. Part 7: Drawings. This section should include all drawings, schematics, etc.

8. Part 8: Commissioning data. This section should include the results of all commissioning works and any remedial works carried out.

9. Part 9: Operation. This section should be used to provide instructions for the safe and efficient operation of the installation, under both normal conditions and emergency conditions, cross-referring to manufacturers' information where appropriate.

10. Part 10: Maintenance. This section should be used to provide schedules for all preventative maintenance, such as inspections, tests, adjustments/re-calibration and lubrication, that are to be carried out on the installation and its component parts.

11. Part 11: Spare parts and special tools. This section should be used to include the following information:

    (i) a list identifying replaceable assemblies, sub-assemblies, and components

    (ii) identification of manufacturers' recommended replacement periods for components that have a low mean time between failures (MTBF), e.g. cooling fans and filters

    (iii)a recommended list of spare parts to be held on site, if components are not readily available

    (iv) information relating to the diagnosis of faults and their correction, where this is not obvious

    (v) where special tools not normally carried by service personnel are required, information on where they can be obtained.

12. Part 12: Disposal. This section should be used to provide details of how the installation is to be de-commissioned, de-constructed and disposed of safely.

The user instructions and drawings prepared by the installer and approved by the designer should be retained with the logbook. If any automatic testing device is installed necessary information on its operation must be included in the instructions.

The logbook is intended to be used for all safety record-keeping, including that of the fire detection and alarm system, in particular all servicing, defects reported, repairs, operation of detectors and alarms including false alarms, attendance of fire brigade etc.

It should also record details of replacement of components, such as luminaires or batteries.

Again it was usual for such logbooks to be paper based but now that digital storage and information systems are normal the logbook can be computerized as long as the occupier/dutyholder and the maintenance staff can access and read the logbook and add entries and amend any details.

6.7 Care of batteries during installation and use

Repeated disconnection of emergency lighting during electrical installation can damage lamps and batteries. So far as is practicable, connection to the emergency lighting should be made only when supplies are unlikely to be regularly disconnected.

Batteries require special safety measures as effectively they cannot be isolated and a connected string of batteries can provide a dangerous voltage. Batteries also contain acid or other dangerous chemicals and must be disposed of correctly at the end of their useful life in accordance with environmental and disposal legislation (see Section 6.12).

Batteries are also sensitive to temperature and operating at temperatures outside their specified operating range can seriously shorten their life expectancy. They can also produce hazardous gasses on charging and discharging and these must be vented safely.

The battery manufacturers operating and safety instructions must be followed.

6.8 Servicing

(Clause 13 of BS 5266-1 and 7 of BS 5266-8)

6.8.1 Supervision

The Electricity at Work Regulations require all persons carrying out electrical work to be competent or properly supervised and BS 5266 requires the owner/occupier of the premises (dutyholder) to appoint a competent person to supervise servicing. *The Regulatory Reform (Fire Safety) Order 2005* or other local legislation puts responsibilities on the responsible person (usually the employer or dutyholder) for maintenance of the fire alarm and fire safety system, including escape lighting. See Regulations 17 and 3 of the Order on the next page.

17. Maintenance

(1) Where necessary in order to safeguard the safety of relevant persons the responsible person must ensure that the premises and any facilities, equipment and devices provided in respect of the premises under this Order or, subject to paragraph (6), under any other enactment, including any enactment repealed or revoked by this Order, are subject to a suitable system of maintenance and are maintained in an efficient state, in efficient working order and in good repair.

3. Meaning of 'responsible person'

In this Order 'responsible person' means –

(a) in relation to a workplace, the employer, if the workplace is to any extent under his control;

(b) in relation to any premises not falling within paragraph (a) –

(i) the person who has control of the premises (as occupier or otherwise) in connection with the carrying on by him of a trade, business or other undertaking (for profit or not); or

(ii) the owner, where the person in control of the premises does not have control in connection with the carrying on by that person of a trade, business or other undertaking.

The Fire (Scotland) Act 2005 places similar duties for maintenance on employers to employees in respect that they all ensure, so far is reasonably practicable, the safety of the employer's employees in respect of harm caused by fire in the workplace.

Note: See also the definition of Responsible Person given in Clause 3.16 of BS 5266-1, but the standard is not a statutory requirement.

6.8.2 Central batteries

(Clause 12 of BS 5266-1)

The manufacturer's instructions are to be followed. Maintenance includes:

(a) keeping batteries and their terminals clean;
(b) checking for leaks; and
(c) topping up the electrolyte.

Care must be taken that any replacement battery is compatible with the battery charger so that charging rates and times are appropriate; similarly, replacement chargers must be compatible with the batteries. Motor vehicle batteries are unlikely to be suitable.

Generally central batteries can last 15 to 20 years, while self-contained luminaire batteries should be of a type having a life expectancy of at least 4 years. This life expectancy of course requires the battery area ambient temperature to be maintained within the manufacturers specified limits. High ambient temperatures can significantly reduce the life of a battery.

6.8.3 Generators

(Clause 12 of BS 5266-1)

The manufacturer's instructions must be obtained, retained and followed. Poor maintenance will often result in failure of the emergency lighting.

Generators and control systems require regular maintenance, including test runs on load. See BS 7698-12:1998 (ISO 8528-12:1997) *Emergency power supply to safety services* and the guidance material published by The Association of Manufacturers and Suppliers of Power Systems (AMPS) for further details.

(Clause 12 of BS 5266-1 and 7.2 of BS 5266-8)

6.9 Routine inspection and testing

Emergency lighting systems should be inspected and tested at regular intervals to confirm that they are properly operational. The testing may be carried out manually or, if this is not practicable, by using an automatic system.

BS 5266-1 advises that guidance on conducting routine tests will have been given to the user as part of the on-site training and handover procedure. Inspections and tests must be undertaken by a competent person or a person properly supervised by a competent person. The provision of an adequate and properly operational emergency lighting system where necessary is a part of the statutory fire safety legislation requirements as discussed in Chapter 1 of this Guide.

Note: When automatic testing is incorporated in the system, the test results shall be recorded monthly. For manual systems the frequency shall be as described below.

(Clause 7.2.2 of BS 5266-8)

6.9.1 Daily visual operational check

Visual inspection of central control power supply indicators to check that the system is in an operational and ready condition. This does not require any testing and may be a duty of security patrols etc.

All members of staff should be encouraged to report any damage they notice at any time.

(Clause 7.2.3 of BS 5266-8)

6.9.2 Monthly installation functional check

If automatic testing is incorporated, the results of these automated tests must be recorded and reviewed, and retained for any future inspection. For manual systems the following checks shall be carried out.

(a) By simulation of failure of the supply to the normal lighting, switch on in the emergency mode each luminaire and each internally illuminated sign for a period sufficient only to check that each lamp is illuminated.
(**Note:** The period of simulated failure needs to be sufficient to check that all the luminaires are illuminated but not such as to impose a significant drain on the batteries.)

(b) At the end of this test period, the supply to the normal lighting is restored and every indicator lamp or device checked to ensure that it is showing that the normal supply has been restored.

(c) Any faults identified and any failed lamps must be repaired or replaced immediately.

(d) For central supply systems, correct operation of the system monitors is checked. For standby supplies to central systems such as batteries and generators, checks are made on the standby supply as recommended by the supplier. Any failure of the generators to start must be rectified immediately.

(e) The tests and results are recorded in the logbook.

(Clause 7.2.4 of BS 5266-8)

6.9.3 Annually

Where automatic testing facilities are installed the results of the full rated duration tests shall be recorded. For manual systems the following checks shall be carried out.

(a) By simulation of failure of the supply to the normal lighting, switch on in the emergency mode each luminaire and each internally illuminated sign for its full rated duration. Any self-contained luminaires that fail to reach their rated duration are to be repaired or replaced (it will usually be a battery failure).

> **Note:** This full duration test must be carried out when it is convenient and safe to do so (for example, over the weekend) as after the test, the emergency lighting will not be available until it has recharged.

(b) At the end of this test period, the supply to the normal lighting is restored and every indicator lamp or device is checked to ensure that it is showing that the normal supply has been restored. The charging arrangements are checked insofar as it is possible to do so.

(c) For central supply systems, correct operation of the system monitors is checked.

(d) For standby supplies to central systems such as batteries and generators, checks are made on the standby supply during and after the test run as recommended by the manufacturer in the operation and maintenance documentation.

(e) For generating sets the requirements of ISO 8528-12 are to be met (it would be wise to have the annual service of a generator just before this annual emergency lighting inspection and test).

(f) The tests and results are recorded in the logbook.

(g) A certificate is supplied to the person ordering the work and/or responsible for the safety systems of the premises when the inspection and testing are carried out by an outside body.

> **Note:** The fire risk assessment required by regulation 9 of the *Regulatory Reform (Fire Safety) Order* may result in a decision to increase maintenance frequencies. *The Fire (Scotland) Act 2005* places similar duties for maintenance on employers and dutyholders.

6.10 Certificates

(Part 6 of
BS 7671,
Clause 6.1
of
BS 5266-8
and
Clause 11
of
BS 5266-1)

6.10.1 Installation Completion Certification

On completion of the installation of an emergency lighting system or an alteration or amendment a commissioning test should be performed to check that the system meets the design requirements and an Emergency Lighting Completion Certificate, or for small new installations (as defined in BS 5266-1) of up to 25 self-contained luminaires an Emergency Lighting Small New Installation Completion Certificate, together with schedules and a logbook should be supplied to the person ordering the work, for forwarding to the occupier or owner of the premises (see Annex H for the model Emergency Lighting Completion Certificates and Annex J for the minimum amount of information a logbook should contain).

A copy of this certificate may be required by the enforcing authority.

Where a logbook already exists for an existing installation which is to be modified the information it contains should be integrated into the new logbook. Where the original Emergency Lighting Completion Certificate is not available, an inspection of the installation is recommended and a certificate for 'Verification of existing installations' should be issued (see Annex K).

A BS 7671 Electrical Installation Certificate and schedules of inspection and test results are required for the wiring associated with the emergency lighting installation and separate certificates for the emergency lighting and any equipment (battery systems etc.) installed (see Annex A to this chapter).

Copies of the certificates along with any operation and maintenance documentation and drawings should be noted in the logbook and copies of the documents retained in the system maintenance files for future reference.

(Part 6 of
BS 7671 and
Clause 6.2 of
BS 5266-8)

6.10.2 Periodic inspection and test certificate

On completion of the yearly inspection and test, an Emergency Lighting Periodic Inspection and Test Certificate (with an Electrical Installation Condition Report if required) shall be provided to the person ordering the work for retention on the premises and copies retained in the emergency lighting logbook for future reference.

Many organizations provide preprinted forms and electronic forms. All are probably suitable but need to be checked against the model forms from BS 5266-1. See Annex D to this Chapter.

Note: Chapter 65 of BS 7671 recommends that periodic inspection and testing of the electrical installation shall be carried out to determine so far as is reasonably practicable whether the electrical installation is in a satisfactory condition for continued service at regular intervals (for detailed information, refer to the IET Guidance Note 3 *Inspection and Testing*). This should include all wiring and equipment associated with the emergency lighting system and the outcome of this inspection and test should be recorded on an Electrical Installation Condition Report and issued to the person ordering the inspection and testing.

6.11 Automatic testing systems (ATS)

BS EN 62034 *Automatic test systems for battery powered emergency escape lighting.*

6.11.1 Classification of ATS types

The BS EN standard classifies automatic testing systems (ATS) as follows.

Type S
This is a stand-alone ATS consisting of a self-contained luminaire with a built-in testing facility that provides a local indication of the condition of the luminaire, but still requires all luminaires to be manually inspected, with a manual record made of the information indicated by the luminaires.

Type P
The emergency luminaires are monitored and their condition is indicated by a test facility that collects and displays the results of the tests, but requires manual recording of information on the tests.

Type ER
As type P, but the test facility collects results, and data is recorded and logged by the ATS.

Type PER
As types P or ER, but with a collated fault indicator that automatically gives remote indication of failure of any of the luminaires that have been tested.

Type PERC
As type PER, but with the additional features of a central controller, for setting parameters, configuration of the system and the central controlled initialization of the test and where the date, time and duration of the test is defined by the central controller.

The introduction to the emergency lighting standard advises that automatic test systems will still require manual intervention to correct faults when they are identified, and procedures should be put in place for such intervention.

A visual check of system components and indicators should be included in the routine of safety staff. This check should be made regularly to ensure that emergency luminaires remain present and intact.

6.12 The Waste Electrical and Electronic Equipment (WEEE) Directive

The Waste Electrical and Electronic Equipment (WEEE) Directive came into force in January 2007 and aims to both reduce the amount of WEEE being produced and encourage everyone to reuse, recycle and recover it. The WEEE Directive also aims to improve the environmental performance of businesses that manufacture, supply, use, recycle or recover electrical and electronic equipment.

The WEEE legislation introduced new responsibilities for businesses and other non-household users of electrical and electronic equipment (EEE). This includes businesses, schools, hospitals and government agencies, when they dispose of their electrical waste. These organizations need to ensure that all separately collected WEEE is treated and recycled. Whether the business or the producer of the EEE pays for this, depends on the circumstances. Gas discharge lamps and LEDs and the luminaires are within the scope of the WEEE Directive.

Factsheets are available from the Environment Agency website.

6.13 Guidance

The guidance given in this chapter is not exhaustive. Further guidance is available in the publications of BSRIA, CIBSE and local authorities, for example:

(a) BSRIA – Illustrated guide to electrical building services BG32/2014;

(b) BSRIA – *Handover, O&M Manuals and Project Feedback: A toolkit for designers and contractors BG1/2007;* and

(c) Society of Light and Lighting (CIBSE) Lighting Guide 12: *Emergency lighting (2015).*

Annex A – Model completion certificate

(from Annex H of BS 5266-1)

This certificate is required when installing a complete system, part of a system or carrying out an installation modification.

The different parties to the installation – Client, Designer, Installer, Commissioning Engineer etc. – may be required to sign this completion certificate – if the design, installation and verification etc. was undertaken by different organizations for example. It is unfortunate that many installation contractors do not properly understand this and sign the 'design' section – even when design (and liability for such) has been by other parties.

A1   Emergency lighting completion certificate

▼ A.1 Emergency lighting completion certificate (Fig H.1 of BS 5266-1:2016)

Serial Number:..................

EMERGENCY LIGHTING COMPLETION CERTIFICATE

For New Installations

Occupier/owner..

Address of premises ....

....

Declaration of Conformity

In consequence of acceptance of the appended declarations, I/we* hereby declare that the emergency lighting system installation, or part thereof, at the above premises conforms, to the best of my/our* knowledge and belief, to the appropriate recommendations given in BS 5266-1:2016, *Emergency lighting – Part 1: Code of practice for the emergency lighting of premises,* BS EN 1838:2013 *Lighting applications – Emergency lighting* and BS EN 50172:2004, *Emergency escape lighting systems,* as set out in the accompanying declarations, except as stated below/overleaf.

* Delete as appropriate.

Signed, on behalf of owner/occupier..

Name...

Deviations from standards

Declaration (Design, installation or verification)	Clause number	Details of deviation

This Certificate is only valid when accompanied by current:

a)   Signed declaration(s) of design, installation and verification, as applicable (see overleaf).

b)   Photometric design data. This can be in any of the following formats but in all cases appropriate de-rating factors must be used and identified to meet worst case requirements.

   • Authenticated spacing data such as ICEL 1001 registered tables**.

   • Calculations as detailed in BS 5266-1:2016, Annex D, and CIBSE/SLL Guide LG12***.

   • Appropriate computer print-out of results.

c)   Test log book.

**Available from Industry Committee for Emergency Lighting, Stafford Park 7, Telford TF3 3BQ.
***Available from Chartered Institution of Building Services Engineers, Delta House, 222 Balham High Road, London SW12 9BS.

A2   Design: declaration of conformity

▼ A.2 Completion certificate - Design - Declaration of conformity Part 1
(Fig H.2 of BS 5266-1:2016)

Serial Number:.................

Design – Declaration of conformity

BS 5266-1: 2016 clause ref.	Recommendations	System conforms (if NO, record a deviation)		
		YES	NO	N/A
4.2	**D1** Accurate plans available showing escape routes, fire alarm control panel, call points and fire extinguishers			
5.2.9	**D2** Escape route signs in accordance with BS EN ISO 7010 and BS 5499-4 and other safety signs in accordance with BS EN ISO 7010 and BS 5499-10, clearly identifiable and adequately illuminated			
6.7	**D3** The luminaires conform to BS EN 60598-2-22			
5.2.8.1	**D4** Luminaires located at following positions: NOTE Near means within 2 m horizontally. a) At each exit door intended to be used in an emergency b) Near stairs so each tread receives direct light c) Near any other change in level d) Externally illuminated escape route signs, escape route direction signs and other safety signs needing to be illuminated under emergency lighting conditions e) At each change of direction f) At intersections of corridors g) Near to each final exit and outside the building to a place of safety h) Near each first aid post i) Near each piece offire-fighting equipment and call point j) Near escape equipment provided for disabled people k) Near refuges and call points, including include two-way communication systems and disabled toilet alarm call position l) Near manual release controls provided to release electronically locked doors			
6.3	**D5** Each room (open area) and escape route has visible light from at least two emergency luminaires			
5.2.8	**D6** Additional emergency lighting provided where needed to illuminate:			
5.2.8.3	a) Evacuation lift cars			
5.2.8.4	b) Moving stair ways and walkways			
5.2.8.5	c) Toilet facilities larger than 8 m^2 floor area or without borrowed light, and those for use by disabled people			
5.2.8.6	d) Motor generator, control, plant and switch rooms			
5.2.8.7	e) Covered car parks			
6.7.3	**D7** Design duration adequate for the application			
10.6; 10.7; Clause **11**	**D8** Operation and maintenance instructions and a suitable log book produced for retention and use by the building occupier			
5.2.5; 5.2.6; 5.2.7	**D9** At least the minimum illuminance provided for escape routes, open areas and high risk task areas			
5.3.2	**D10** At least the minimum illuminance provided for emergency safety lighting			

▼ A.2 Completion certificate - Design - Declaration of conformity Part 2
(Fig H.2 of BS 5266-1:2016)

Deviations from standards (to be entered on Completion Certificate)	
Clause number	**Details of deviation**

Name of competent person making the design declaration of conformity (please print)

....

Signature of competent person ....

For and on behalf of .... Date...

A3   Installation: declaration of conformity

▼ A.3 Completion certificate – Installation – Declaration of conformity
(Fig H.3 of BS 5266-1:2016)

Serial Number:................

Installation – Declaration of conformity

BS 5266-1: 2016 clause ref.	Recommendations	System conforms (if NO, record a deviation)		
		YES	NO	N/A
Clause 6	**IN1** The system installed conforms to the agreed design			
6.1	**IN2** All non-maintained luminaires fed or controlled by the final circuit supply of their local normal mains lighting			
6.4	**IN3** Luminaires mounted at least 2 m above the floor			
6.4	**IN4** Luminaires mounted at a suitable height to avoid being located in smoke reservoirs or other likely area of smoke accumulation			
5.2.9 5.2.9.1 5.2.9.2	**IN5** Safety signs provided as follows: a) Escape route signs in accordance with BS EN ISO 7010 and BS 5499-4, adequately illuminated and identifiable b) Other safety signs in accordance with BS EN ISO 7010 and BS 5499-10, adequately illuminated and identifiable			
8.2	**IN6** The wiring of central power systems has adequate fire protection and is appropriately sized			
8.3.5	**IN7** Output voltage range of the central power system is compatible with the supply voltage range of the luminaires, taking into account supply cable voltage drop			
8.2.12	**IN8** All plugs and sockets protected against unauthorized use			
8.3.3	**IN9** The system has suitable and appropriate testing facilities for the specific site			
Clause 11	**IN10** The equipment manufacturers' installation and verification procedures satisfactorily completed			
Clause 8	**IN11** The system conforms to BS 7671			

Deviations from standards
(to be entered on Completion Certificate)

Clause number	Details of deviation

Name of competent person making the installation declaration of conformity (please print)

....

Signature of competent person ....

For and on behalf of .... Date..

A4 Verification: declaration of conformity

▽ A.4 Completion certificate – Verification – Declaration of conformity Part 1
(Fig H.4 of BS 5266-1:2016)

Serial Number:..................

Verification – Declaration of conformity

BS 5266-1: 2016 clause ref.	Recommendations	System conforms (if NO, record a deviation)		
		YES	NO	N/A
4.2	**V1** Plans available and correct			
8.3.3	**V2** System has a suitable test facility for the application			
5.2.9	**V3** All escape route safety signs and fire-fighting equipment location signs, and other safety signs identified from risk assessment, visible with the normal lighting extinguished			
Clause 5	**V4** Luminaires correctly positioned and oriented as shown on the plans			
6.7.1	**V5** Luminaires conform to BS EN 60598-2-22			
6.7.1	**V6** Luminaires have an appropriate category of protection against ingress of moisture or foreign bodies for their location as specified in the system design			
Clause 12	**V7** Luminaires tested and found to operate for their full rated duration			
Clause 12	**V8** Under test conditions, adequate illumination provided for safe movement on the escape route and the open areas, paths under emergency safety lighting, and operations within high risk task areas NOTE This can be checked by visual inspection and checking that the illumination from the luminaires is not obscured and that minimum design spacings have been met.			
Clause 12	**V9** After test, the charging indicators operate correctly			
8.2	**V10** Fire protection of central wiring systems satisfactory			
8.2.6	**V11** Emergency circuits correctly segregated from other supplies			
10.6; 10.7; Clause 11	**V12** Operation and maintenance instructions together with a suitable log book showing a satisfactory verification test provided for retention and use by the building occupier			

Additional recommendations for verification of an existing installation

10.7 and Clause 12	**V13** Building occupier and their staff trained on suitable maintenance, testing and operating procedures, or a suitable maintenance contract agreed			
Clause 11	**V14** Test records in the log book complete and satisfactory			
Clause 12	**V15** Luminaires clean and undamaged with lamps in good condition			
Clause 6	**V16** Original design still valid NOTE If the original design is not available this needs to be recorded as a deviation.			

▼ A.4 Completion certificate – Verification – Declaration of conformity Part 2
(Fig H.4 of BS 5266-1:2016)

Deviations from standards (to be entered on Completion Certificate)	
Clause number	**Details of deviation**

Name of competent person making the verification declaration of conformity (please print)

....

Signature of competent person ....

For and on behalf of .... Date.......................................

Annex B – Model certificate for completion of small new installations

(The definition of 'small installation' under BS 5266-1 is one with up to 25 self-contained luminaires.)

B1   General declaration

▼ B.1 Certificate for completion of small new installations – General declaration (Fig I.1 of BS 5266-1:2016)

Serial Number:.................

EMERGENCY LIGHTING SMALL NEW INSTALLATIONS COMPLETION CERTIFICATE

For Small New Installations up to 25 self-contained luminaires

Occupier/owner...

Address of premises ....

....

Declaration of Conformity

In consequence of acceptance of the appended declarations, I/we* hereby declare that the emergency lighting system installation, or part thereof, at the above premises conforms, to the best of my/our* knowledge and belief, to the appropriate recommendations given in BS 5266-1:2016, *Emergency lighting – Part 1: Code of practice for the emergency lighting of premises*, BS EN 1838:2013 *Lighting applications – Emergency lighting* and BS EN 50172:2004, *Emergency escape lighting systems*, as set out in the accompanying declarations, except as stated below/overleaf.

* Delete as appropriate.

Signed, on behalf of owner/occupier...

Name..

Deviations from standards

Declaration (Design, installation or verification)	Clause number	Details of deviation

This Certificate is only valid when accompanied by current:

a)   Signed declaration (see overleaf).

b)   Photometric design data. This can be in any of the following formats but in all cases appropriate de-rating factors must be used and identified to meet worst case requirements.

   • Authenticated spacing data such as ICEL 1001 registered tables**.

   • Calculations as detailed in BS 5266-1:2016, Annex D, and CIBSE/SLL Guide LG12***.

   • Appropriate computer print-out of results.

c)   Test log book.

**Available from Industry Committee for Emergency Lighting, Stafford Park 7, Telford TF3 3BQ.
***Available from Chartered Institution of Building Services Engineers, Delta House, 222 Balham High Road, London SW12 9BS.

B2 Design, installation and verification: declaration of conformity

▼ B.2 Certificate for completion of small new installations – Declaration of conformity Part 1 (Fig I.2 of BS 5266-1:2016)

Serial Number:................

Site address			Responsible person			
BS 5266-1: 2016 clause ref.	Competent person function D-Designer, I-Installer, V-Verifier		Inspection Date			
	D,I,V	**Check of categories and documentation**		**YES**	**NO**	**N/A**
4.2	D,V	Are plans of the system available and correct?				
6.7	D,V	Has the system been designed for the correct mode of operation category?				
6.7	D,V	Has the system been designed for the correct emergency duration period?				
Clause 11	D,V	Is a completion certificate available with photometric design data?				
Clause 11	D,I,V	Is a test log book available and are the entries up to date?				
		Check of design				
4.1; 5.2.8	D,I,V	Are the correct areas of the premises covered to meet the risk assessment?				
5.2.8	D,I,V	Are all hazards identified by the risk assessment covered?				
5.2.8	D,I,V	Are there luminaires sited at the "points of emphasis"?				
5.2.2	D,I,V	Is the spacing between luminaires compliant with authenticated spacing or design data?				
5.2.9	D,I,V	Are the emergency exit signs and escape route direction signs correct and the locations of other safety signs to be illuminated under emergency conditions identified?				
6.1	D,I,V	Do all non-maintained luminaires operate on local final circuit failure?				
6.3	D,V	Is there illumination from at least two luminaires in each section of the escape route?				
6.4	I,V	Are luminaires at least 2 m above floor and avoiding smoke reservoirs?				
5.2.8.5; 5.2.8.6	D,V	Are additional luminaires located to cover toilets, lifts, plant rooms, etc.?				
		Check of the quality of the system components and installation				
6.7	D,I,V	Do the luminaires conform to BS EN 60598-2-22?				
6.7	D,I,V	Do any converted luminaires conform to BS EN 60598-2-22?				
6.7	D,I,V	Do luminaires have a suitable degree of protection for their location?				
Clause 8	I,V	Does the installation conform to the good practice defined in BS 7671?				
8.2.12	D,I,V	Are any plugs or sockets protected against unauthorized use?				

B2 Design, installation and verification: declaration of conformity *continued*

▼ B.2 Certificate for completion of small new installations – Declaration of conformity Part 2 (Fig I.2 of BS 5266-1:2016)

		Test facilities	YES	NO	N/A
8.3.3	D,V,I	Are the test facilities suitable to test function and duration?			
8.3.3	D,IV	Are the test facilities safe to operate and do not isolate a required service?			
8.3.3	D,IV	Are the test facilities clearly marked with their function?			
8.3.3	D,IV	If an automatic test system is installed, does it conform to BS EN 62034?			
10.7	D,V	Is the responsible person trained and able to operate the test facilities and record the test results correctly?			
		Final acceptance to be conducted at completion			
Clause **12**	D,IV	Does the system operate correctly when tested?			
10.7	D,IV	Has adequate documentation been provided to the user?			
10.7	D,IV	Is the user aware of action they should take in the event of a test failure?			

Action recommended or deviation to be reported:

Name of competent person making the declaration of conformity (please print)

....

Signature of competent person ....

For and on behalf of .... Date............................

Annex C – Model certificate for verification of existing installations

C1 Certificate for verification of existing installations

▼ C.1 Certificate for completion of existing installations - General declaration (K.1 of BS 5266-1:2016)

Serial Number:.................

EMERGENCY LIGHTING EXISTING SITE COMPLIANCE CERTIFICATE

For Verification of Existing Installations

Occupier/owner...

Address of premises ....

....

Declaration of Conformity

In consequence of acceptance of the appended declarations, I/we* hereby declare that the emergency lighting system installation, or part thereof, at the above premises conforms, to the best of my/our* knowledge and belief, to the appropriate recommendations given in BS 5266-1:2016, *Emergency lighting – Part 1: Code of practice for the emergency lighting of premises*, BS EN 1838:2013 *Lighting applications – Emergency lighting* and BS EN 50172:2004, *Emergency escape lighting systems*, as set out in the accompanying declarations, except as stated below/overleaf.

* Delete as appropriate.

Signed, on behalf of owner/occupier...

Name..

Deviations from standards

Declaration (Design, installation or verification)	Clause number	Details of deviation

This Certificate is only valid when accompanied by current:

a) Signed checklist and report, as applicable (see overleaf).

b) Photometric design data. This can be in any of the following formats but in all cases appropriate de-rating factors must be used and identified to meet worst case requirements.

· Authenticated spacing data such as ICEL 1001 registered tables**.

· Calculations as detailed in BS 5266-1:2016, Annex D, and CIBSE/SLL Guide LG12***.

· Appropriate computer print-out of results.

· Site test light readings.

c) Test log book.

**Available from Industry Committee for Emergency Lighting, Stafford Park 7, Telford TF3 3BQ.
***Available from Chartered Institution of Building Services Engineers, Delta House, 222 Balham High Road, London SW12 9BS.

C2 Certificate for verification of existing installations - Checklist and report

▽ C.2 Certificate for verification of existing installations - Checklist and report Part 1 (K.2 of BS 5266-1:2016)

Serial Number:..................

Site address			Responsible person			
BS 5266-1: 2016 clause ref.	Competent person function D-Designer, I-Installer, V-Verifier		Inspection Date			
	D,I,V	**Check of categories and documentation**		**YES**	**NO**	**N/A**
4.2	D,V	Are plans of the system available and correct?				
6.7	D,V	Has the system been designed for the correct mode of operation category?				
6.7	D,V	Has the system been designed for the correct emergency duration period?				
Clause 11	D,V	Is a completion certificate available with photometric design data?				
Clause 11	D,I,V	Is a test log book available and are the entries up to date?				
		Check of design				
4.1; 5.2.8	D,I,V	Are the correct areas of the premises covered to meet the risk assessment?				
5.2.8	D,I,V	Are all hazards identified by the risk assessment covered?				
5.2.8	D,I,V	Are there luminaires sited at the "points of emphasis"?				
5.2.2	D,I,V	Is the spacing between luminaires compliant with authenticated spacing or design data?				
10.3	D,I,V	If authenticated spacing data is not available for existing installations, are estimates attached and acceptable?				
5.2.9	D,I,V	Are the emergency exit signs and escape route direction signs correct and the locations of other safety signs to be illuminated under emergency conditions identified?				
6.1	D,I,V	Do all non-maintained luminaires operate on local final circuit failure?				
6.3	D,V	Is there illumination from at least two luminaires in each section of the escape route?				
6.4	I,V	Are luminaires at least 2 m above floor and avoiding smoke reservoirs?				
5.2.8.5; 5.2.8.6	D,V	Are additional luminaires located to cover toilets, lifts, plant rooms, etc.?				
		Check of the quality of the system components and installation				
6.7	D,I,V	Do the luminaires conform to BS EN 60598-2-22?				
6.7	D,I,V	Do any converted luminaires conform to BS EN 60598-2-22?				
6.7	D,I,V	Do luminaires have a suitable degree of protection for their location?				
Clause 8	I,V	Does the installation conform to the good practice defined in BS 7671?				
8.2.1	D,I,V	For centrally powered systems, is the wiring fire-resistant?				
8.2.12	D,I,V	Are any plugs or sockets protected against unauthorized use?				
7.2	D,I,V	If a central power supply unit is used, does it conform to BS EN 50171?				

C3   Certificate for verification of existing installations - Checklist and report *continued*

▼ C.3 Certificate for verification of existing installations - Checklist and report Part 2 (K.2 of BS 5266-1:2016)

		Test facilities	YES	NO	N/A
8.3.3	D,V,I	Are the test facilities suitable to test function and duration?			
8.3.3	D,IV	Are the test facilities safe to operate and do not isolate a required service?			
8.3.3	D,IV	Are the test facilities clearly marked with their function?			
8.3.3	D,IV	If an automatic test system is installed, does it conform to BS EN 62034?			
10.7	D,V	Is the responsible person trained and able to operate the test facilities and record the test results correctly?			
		Final acceptance to be conducted at completion			
Clause **12**	D,IV	Does the system operate correctly when tested?			
10.7	D,IV	Has adequate documentation been provided to the user?			
10.7	D,IV	Is the user aware of action they should take in the event of a test failure?			

Action recommended or deviation to be reported:

Name of competent person completing the checklist and report (please print)

....

Signature of competent person ....

For and on behalf of .... Date................................

Annex D – Model periodic inspection and test certificate

This certificate should be used to report on the condition of an existing emergency lighting installation where there are perhaps no records or previous certificates available. This should usually be done on before any repairs, upgrading or additional works are carried out, or perhaps before a new maintenance contact commences. It is not to be used to record regular maintenance activities and regular testing of an existing emergency lighting system.

D1   Periodic inspection and test certificate

▼ D.1 Emergency lighting inspection and test certificate
(Fig M.1 of BS 5266-1:2016)

Emergency Lighting Inspection and Test Certificate

For systems designed to BS 5266-1 and BS EN 50172/BS 5266-8

WARNING

Full duration tests involve discharging the batteries, so the emergency lighting system will not be fully functional until the batteries have had time to recharge. For this reason, always carry out testing at times of minimal risk, or only test alternate luminaires at any one time.

System manufacturer Contact phone number			
System installer Contact phone number			
Competent person responsible for verification and annual tests			Phone number
Name .... Signature			

Site address

Responsible person			
Date the system was commissioned			
Details of system mode of operation	Non-maintained		
	Non-maintained luminaires, maintained signs		
	Maintained		
	Other		
Duration of system	.... Hours	Is automatic test system fitted?	Y/N

Details of additions or modifications to the system or the premises since original installation

Addition or modification		Date

Action to be taken on finding a failure

- The supplier of the system or a competent person should be contacted to rectify the fault.

- A risk assessment of the failure should be conducted; this should evaluate the people who will be at increased risk and the level of that risk. Based on this data and, if necessary, advice from the Fire Authority, the appropriate action should be taken.

- Action may be:

  To warn occupants to be extra vigilant until the system is rectified

  To initiate extra safety patrol

  To issue torches as a temporary measure

  In a high risk situation, to limit use of all or part of the building

Test programs for identifying early failures can reduce the chances offailure of two adjacent luminaires at the same time.

D2   Emergency lighting inspection and test record

▼ D.2 Emergency lighting inspection and test record (Fig M.2 of BS 5266-1:2016)

Emergency Lighting Inspection and Test Record			Sheet number:	
Site:				
Test types:	C = Commissioning and verification test			
	M = Monthly test (see BS EN 50172:2004/BS 5266-8:2004, **7.2.3**)			
	A = Annual test (see BS EN 50172:2004/BS 5266-8:2004, **7.2.4**)			
Date of test	**Test type**	**Result – Test passed No action needed**	**Result – Test failed**	
			Need for repair of system notified	**Need for safeguarding of premises notified**
		Name	Name	Name
	C			
	M – 1st month			
	M – 2nd month			
	M – 3rd month			
	M – 4th month			
	M – 5th month			
	M – 6th month			
	M – 7th month			
	M – 8th month			
	M – 9th month			
	M – 10th month			
	M – 11th month			
	A – 1st year			
	M – 1st month			
	M – 2nd month			
	M – 3rd month			
	M – 4th month			
	M – 5th month			
	M – 6th month			
	M – 7th month			
	M – 8th month			
	M – 9th month			
	M – 10th month			
	M – 11th month			
	A – 2nd year			
	M – 1st month			
	M – 2nd month			
	M – 3rd month			
	M – 4th month			
	M – 5th month			
	M – 6th month			
	M – 7th month			
	M – 8th month			
	M – 9th month			
	M – 10th month			
	M – 11th month			
	A – 3rd year			

D3 Emergency lighting fault action record

▼ D.3 Emergency lighting fault action record (Fig M.3 of BS 5266-1:2016)

Emergency Lighting Fault Action Record			Sheet number:		
Contact references		Contact name	Phone number		
Equipment supplier:				For replacement parts	
Maintenabce engineer:				Competent person	
Responsible person:				Site control	
Date of failure	Action taken to safeguard the premises (Details and signature)		Action taken to rectify the system (Details and signature)		Date system repaired

Safety signs 7

7.1 Introduction

7.1.1 The Health and Safety (Safety Signs and Signals) Regulations 1996 (SI 1996 No. 341)

Where the risk assessment required by the *Management of Health and Safety at Work Regulations* indicates a need for any safety signs or signals, the *Health and Safety (Safety Signs and Signals) Regulations* require that suitable signs be installed including if necessary standby supplies.

Relevant parts of the *Safety Signs and Signals Regulations* are reproduced in Section 1.5 of Chapter 1.

The Health and Safety Executive have published guidance on the safety signs and signals of the *Health and Safety (Safety Signs and Signals) Regulations* in their publication L64.

In this chapter only the safety signs associated with fire systems and escape lighting recommended in the publication are reproduced but there may well be other safety signs required in a work area and the HSE guidance advises against confusion between different signage systems for different operations.

7.1.2 Approved Document B

(para 5.37 of Vol. 2)

The Approved Document or other relevant local legislation recommends that except in a flat (as opposed to common areas), every escape route other than those in ordinary use should be distinctively and conspicuously marked by emergency exit signs of adequate size, complying with the *Health and Safety (Safety Signs and Signals) Regulations 1996*. Reference is also made to BS 5499-1 (but see Section 2.5.2 of this Guide).

7.2　Format of safety signs

Schedule 1 to the *Health and Safety (Safety Signs and Signals) Regulations 1996* lays down the format of safety signs, including the shape and colour.

The shapes, etc. recommended in the guidance are as follows.

▼ **Figure 7.1** Prohibitory signs: round shape, black pictogram on white background, red edging and diagonal line

▼ **Figure 7.2** Warning signs: triangular shape, black pictogram on a yellow background with black edging

▼ **Figure 7.3** Mandatory signs: round shape, white pictogram on blue background

▼ **Figure 7.4** Emergency escape and first-aid signs: rectangular or square shape, white pictogram on a green background

Note: Escape route Signs designated E001 and E002 in BS EN ISO 7010:2012+A5 should be used with the appropriate directional arrow in accordance with BS 5499-4:2013, Table 1.

▼ **Figure 7.5** Fire-fighting signs: rectangular or square shape, white pictogram on a red background

7.3 Emergency escape and first-aid signs

The signs given in HSE publication L64 *Safety signs and signals* are reproduced below.

▼ **Figure 7.6** Emergency escape

▼ Figure 7.7 Supplementary 'This way'

▼ Figure 7.8 First aid

a First-aid post   **b** Shower   **c** Stretcher

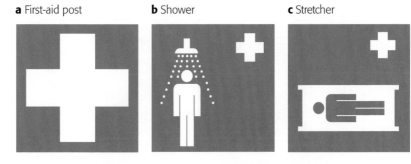

d Eyewash   **e** Emergency telephone

7.4 Fire-fighting signs

▼ **Figure 7.9** Fire-fighting signs

a Fire hose

b Ladder

c Emergency fire telephone

d Fire extinguisher

▼ **Figure 7.10** Supplementary 'This way' signs for fire-fighting equipment

Index